KU-017-766

REEDS MARINA GUIDE 2022

The source directory for all sail & power boat owners

Adlard Coles Nautical,
50 Bedford Square,
London, WC1B 3DP
Tel: 01865 411010
e-mail: info@reedsalmanacs.co.uk
www.reedsalmanacs.co.uk

Cover photo: Neyland Yacht Haven,
Yacht Havens Group 01590 677071
www.yachthavens.com

Section 1

The Marinas and Services Section has been fully updated for the 2022 season. These useful pages provide chartlets and facility details for some 200 marinas around the shores of the UK and Ireland, including the Channel Islands, the perfect complement to any Reeds Nautical Almanac.

Section 2

The Marine Supplies & Services section lists more than 1000 services at coastal and other locations around the British Isles. It provides a quick and easy reference to manufacturers and retailers of equipment, services and supplies both nationally and locally together with emergency services.

Printed by Bell and Bain Ltd, Glagow

Section 1

Section 2

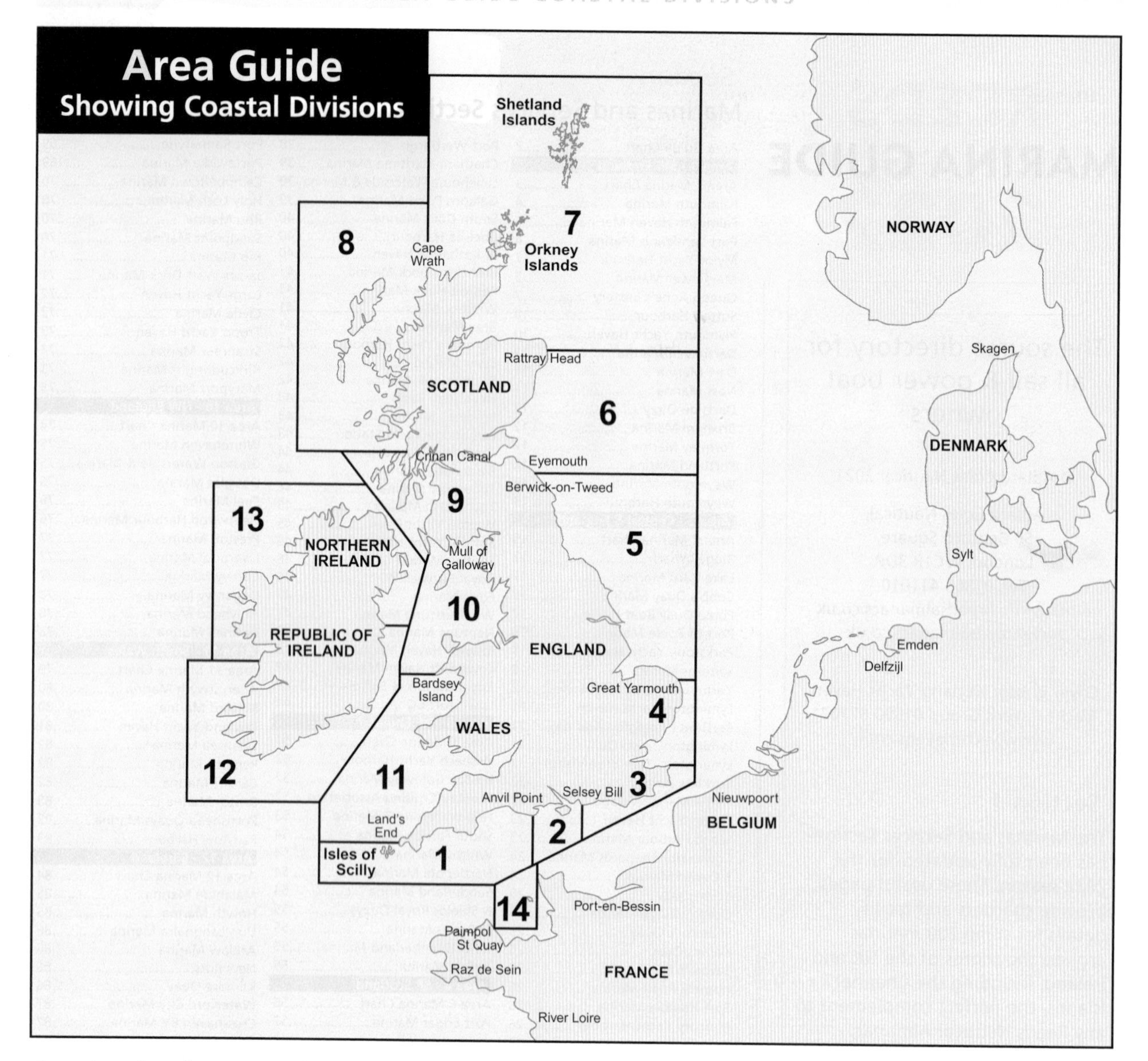
Area Guide
Showing Coastal Divisions
Shetland Islands
7
Orkney Islands
8
Cape Wrath
NORWAY
Skagen
Rattray Head
SCOTLAND
6
DENMARK
Crinan Canal
Eyemouth
Berwick-on-Tweed
9
13
NORTHERN IRELAND
Mull of Galloway
5
Sylt
10
REPUBLIC OF IRELAND
ENGLAND
Emden
Delfzijl
Bardsey Island
Great Yarmouth
4
WALES
12
11
Anvil Point
Selsey Bill
3
Nieuwpoort
BELGIUM
Land's End
2
Isles of Scilly
1
14
Port-en-Bessin
Paimpol
St Quay
Raz de Sein
FRANCE
River Loire

Area 1 .....**South West England**.......... Isles of Scilly to Anvil Point

Area 2 .....**Central Southern England**.......... Anvil Point to Selsey Bill

Area 3 .....**South East England**.......... Selsey Bill to North Foreland

Area 4 .....**East England**.......... North Foreland to Great Yarmouth

Area 5 .....**North East England**.......... Great Yarmouth to Berwick-upon-Tweed

Area 6 .....**South East Scotland**.......... Eyemouth to Rattray Head

Area 7 .....**North East Scotland**.......... Rattray Head to Cape Wrath including Orkney & Shetland Is

Area 8 .....**North West Scotland**.......... Cape Wrath to Crinan Canal

Area 9 .....**South West Scotland**.......... Crinan Canal to Mull of Galloway

Area 10 ...**North West England**.......... Isle of Man & N Wales, Mull of Galloway to Bardsey Is

Area 11 ...**South Wales & Bristol Channel**.......... Bardsey Island to Land's End

Area 12 ...**South Ireland**.......... Malahide, clockwise to Liscannor Bay

Area 13 ...**North Ireland**.......... Liscannor Bay, clockwise to Lambay Island

Area 14 ...**Channel Islands**.......... Guernsey and Jersey

SOUTH WEST ENGLAND – Isles of Scilly to Anvil Point

Key to Marina Plans symbols

Bottled gas	Parking
Chandler	Pub/Restaurant
Disabled facilities	Pump out
Electrical supply	Rigging service
Electrical repairs	Sail repairs
Engine repairs	Shipwright
First Aid	Shop/Supermarket
Fresh Water	Showers
Fuel - Diesel	Slipway
Fuel - Petrol	Toilets
Hardstanding/boatyard	Telephone
Internet Café	Trolleys
Laundry facilities	Visitors berths
Lift-out facilities	Wi-Fi

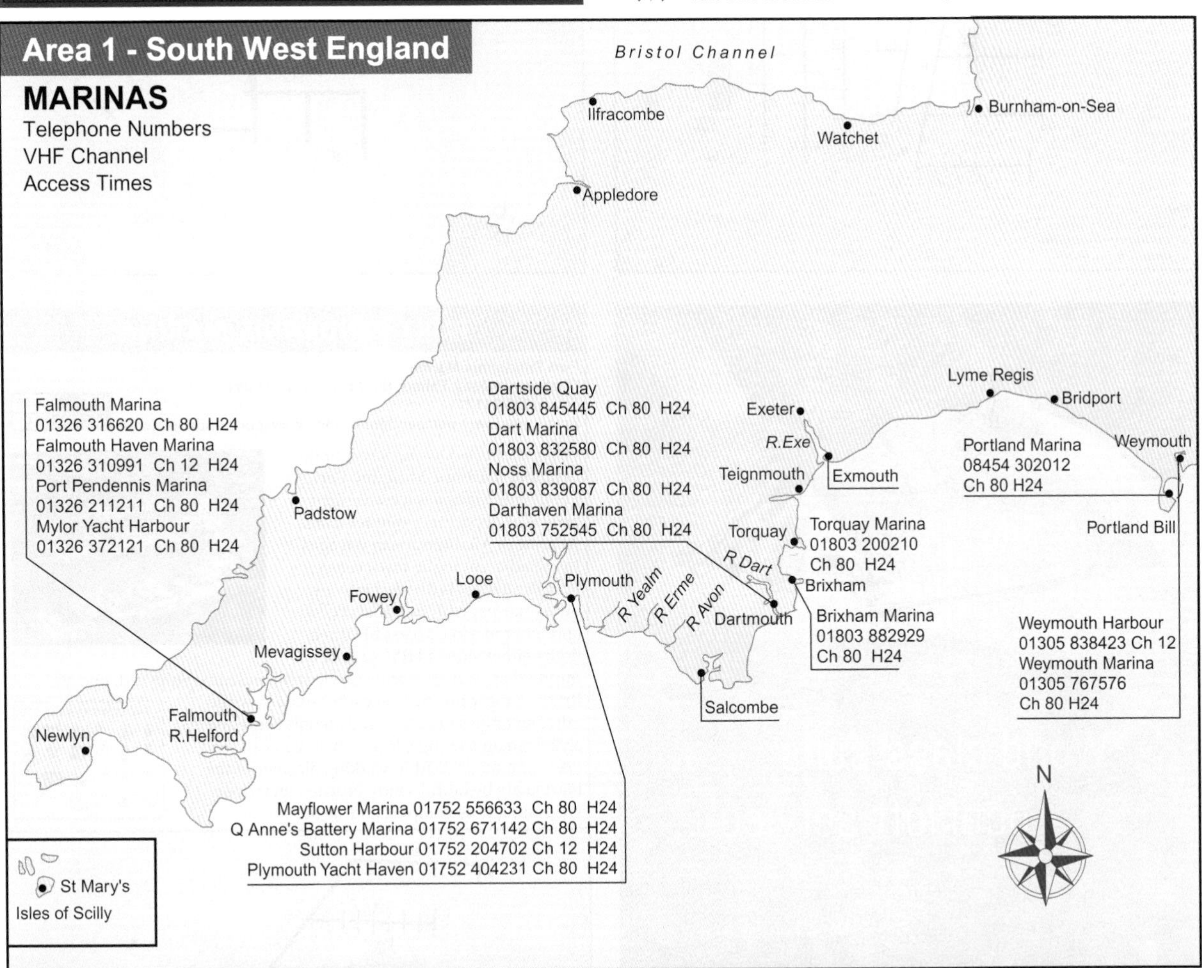

Area 1 - South West England
MARINAS
Telephone Numbers
VHF Channel
Access Times
Bristol Channel
Ilfracombe
Watchet
Burnham-on-Sea
Appledore
Falmouth Marina
01326 316620 Ch 80 H24
Falmouth Haven Marina
01326 310991 Ch 12 H24
Port Pendennis Marina
01326 211211 Ch 80 H24
Mylor Yacht Harbour
01326 372121 Ch 80 H24
Padstow
Dartside Quay
01803 845445 Ch 80 H24
Dart Marina
01803 832580 Ch 80 H24
Noss Marina
01803 839087 Ch 80 H24
Darthaven Marina
01803 752545 Ch 80 H24
Exeter
R.Exe
Teignmouth
Exmouth
Lyme Regis
Bridport
Portland Marina
08454 302012
Ch 80 H24
Weymouth
Portland Bill
Torquay
Torquay Marina
01803 200210
Ch 80 H24
R Dart
Brixham
Brixham Marina
01803 882929
Ch 80 H24
Looe
Fowey
Plymouth
R Yealm
R Erme
R Avon
Dartmouth
Mevagissey
Salcombe
Weymouth Harbour
01305 838423 Ch 12
Weymouth Marina
01305 767576
Ch 80 H24
Falmouth
R.Helford
Newlyn
N
Mayflower Marina 01752 556633 Ch 80 H24
Q Anne's Battery Marina 01752 671142 Ch 80 H24
Sutton Harbour 01752 204702 Ch 12 H24
Plymouth Yacht Haven 01752 404231 Ch 80 H24
St Mary's
Isles of Scilly

FALMOUTH MARINA

Falmouth Marina
North Parade, Falmouth, Cornwall, TR11 2TD
Tel: 01326 316620
Email: falmouth@premiermarinas.com
www.premiermarinas.com

VHF	Ch 80
ACCESS	H24

Falmouth Marina sits within the safe and sheltered waters and outstanding natural beauty of the Fal Estuary, with 24/7 access to the glorious cruising grounds and countryside of Cornwall.

The marina is family-friendly and ideal for boaters of all abilities, with both secure wet and dry stack berthing options. Falmouth provides first-class facilities, along with an array of marine and hospitality services, including a chandlery, brokerage, fully serviced boatyard, bar/restaurant and an onsite hairdresser.

FACILITIES AT A GLANCE

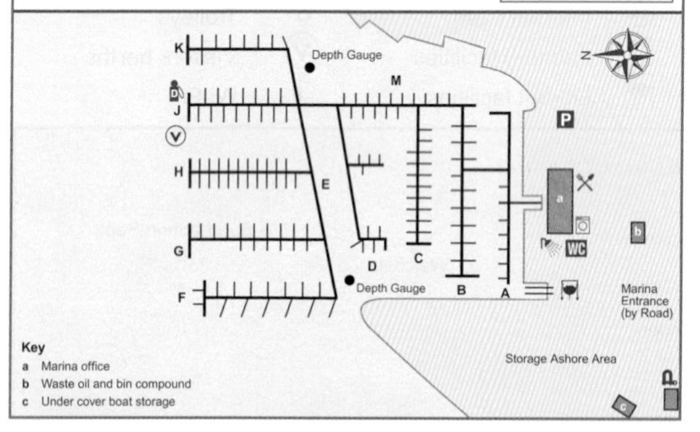
K
J
H
G
F
M
E
D
C
B
A
Depth Gauge
Depth Gauge
Marina Entrance (by Road)
Storage Ashore Area
Key
a Marina office
b Waste oil and bin compound
c Under cover boat storage

FALMOUTH HAVEN MARINA

Falmouth Haven Marina
44 Arwenack Street
Tel: 01326 310991
Email: welcome@falmouthhaven.co.uk

VHF Ch 12
ACCESS H24

Run by Falmouth Harbour Commissioners (FHC), Falmouth Haven Marina has become increasingly popular since its opening in 1982, enjoying close proximity to the amenities and entertainments of Falmouth town centre. Sheltered by a breakwater, Falmouth Haven caters for 50 boats and offers petrol and diesel supplies as well as good shower and laundry facilities.

Falmouth Harbour is considered by some to be the cruising capital of Cornwall and its deep water combined with easily navigable entrance – even in the severest conditions – makes it a favoured destination for visiting yachtsmen.

FACILITIES AT A GLANCE

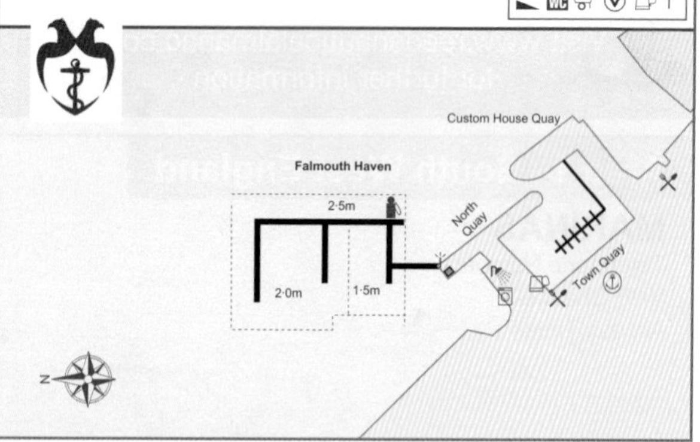
Custom House Quay
Falmouth Haven
2·5m
2·0m
1·5m
North Quay
Town Quay

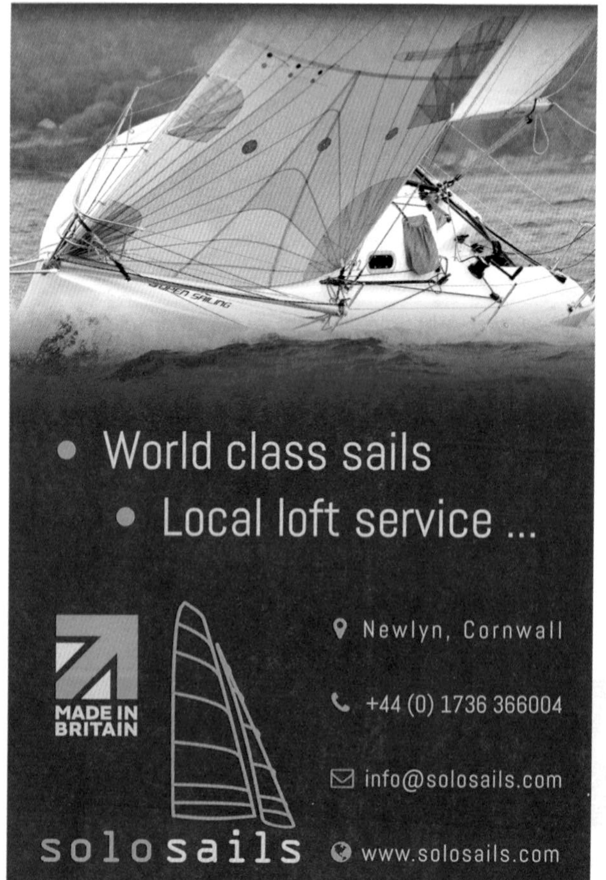
• World class sails
• Local loft service ...
MADE IN BRITAIN
Newlyn, Cornwall
+44 (0) 1736 366004
info@solosails.com
www.solosails.com
solosails

PORT PENDENNIS MARINA

Port Pendennis Marina
Challenger Quay, Falmouth, Cornwall, TR11 3YL
Tel: 01326 211211
Email: marina@.portpendennis.com www.portpendennis.com

VHF	Ch 80
ACCESS	H24

Easily identified by the tower of the National Maritime Museum, Port Pendennis Marina is a convenient arrival or departure point for trans-Atlantic or Mediterranean voyages. Lying adjacent to the town centre, Port Pendennis is divided into an outer marina, with full tidal access, and inner marina, accessible three hours either side of HW. Among its impressive array of marine services is Pendennis Shipyard, one of Britain's most prestigious yacht builders, while other amenities on site include tennis court and a yachtsman's lounge, from where you can send faxes or e-mails. Within walking distance of the marina are beautiful sandy beaches, an indoor swimming pool complex and castle.

FACILITIES AT A GLANCE

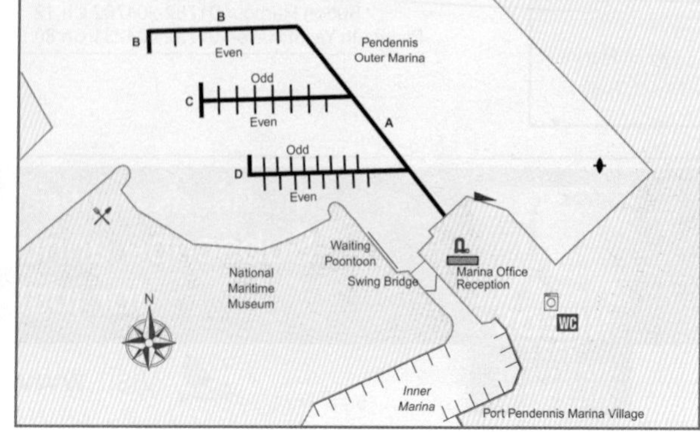
B
B
Even
Pendennis Outer Marina
Odd
C
Even
A
Odd
D
Even
Waiting Pontoon
Swing Bridge
National Maritime Museum
Marina Office Reception
N
Inner Marina
Port Pendennis Marina Village

MYLOR YACHT HARBOUR

Mylor Yacht Harbour Marina
Mylor, Falmouth, Cornwall, TR11 5UF
Tel: 01326 372121
Email: enquiries@mylor.com

VHF	Ch M, 80
ACCESS	H24

Nestled on the western shore of the beautiful Fal Estuary, Mylor Yacht Harbour's stunning marina has 180 berths and is surrounded by 240 swinging moorings. A large dedicated visitor pontoon, easy H24 access and excellent shore-side facilities set within an Area of Outstanding Natural Beauty combine to make Mylor a must-visit cruising destination.

The historic site was founded in 1805 as the Navy's smallest dockyard but is now a thriving yacht harbour and full service boatyard. Superb on site café, restaurant and yacht club all overlook the harbour and scenic coastal footpaths run from the top of the gangway. Falmouth is just 10 minutes away.

FACILITIES AT A GLANCE

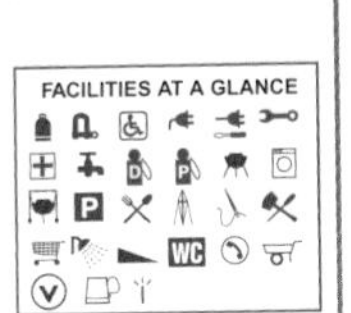

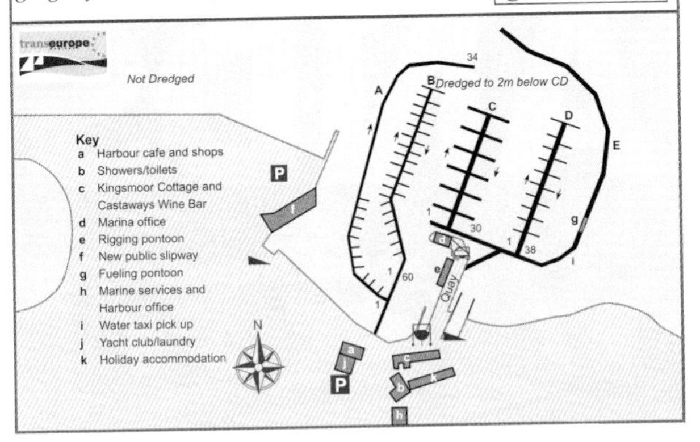
transeurope
Not Dredged
Dredged to 2m below CD
Key
a Harbour cafe and shops
b Showers/toilets
c Kingsmoor Cottage and Castaways Wine Bar
d Marina office
e Rigging pontoon
f New public slipway
g Fueling pontoon
h Marine services and Harbour office
i Water taxi pick up
j Yacht club/laundry
k Holiday accommodation
Quay
N

MAYFLOWER MARINA

Mayflower Marina
Richmond Walk, Plymouth, PL1 4LS
Tel: 01752 556633
Email: info@mayflowermarina.co.uk

VHF	Ch 80
ACCESS	H24

Sitting on the famous Plymouth Hoe, with the Devon coast to the left and the Cornish coast to the right, Mayflower Marina is a friendly, well-run marina. Facilities include 24 hour access to fuel, gas and a launderette, full repair and maintenance services as well as an on site bar and brasserie. The marina is located only a short distance from Plymouth's town centre, where there are regular train services to and from several major towns and cities.

FACILITIES AT A GLANCE

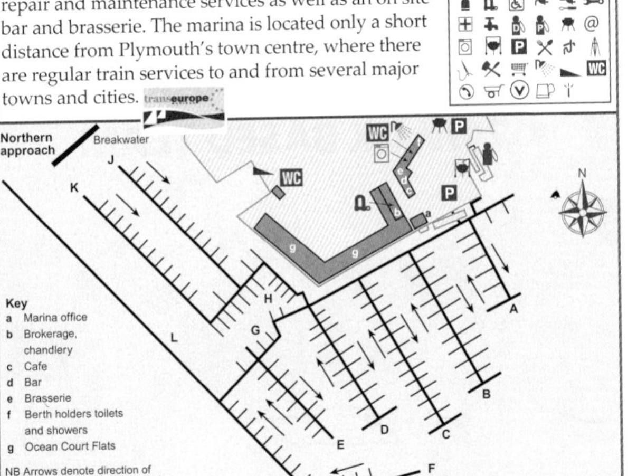
Northern approach
Breakwater
Key
a Marina office
b Brokerage, chandlery
c Cafe
d Bar
e Brasserie
f Berth holders toilets and showers
g Ocean Court Flats
NB Arrows denote direction of numbers low to high. Even numbers on port side, odd to starboard
Southern approach
N

QUEEN ANNE'S BATTERY

Queen Anne's Battery
Plymouth, Devon, PL4 OLP
Tel: 01752 671142
Email: qab@mdlmarinas.co.uk www.queenannesbattery.co.uk

VHF	Ch 80
ACCESS	H24

Plymouth, a vibrant and fast growing city, is home to Queen Anne's Battery, a 235 resident berth marina with outstanding facilities for yachtsmen and motor cruisers alike. Located just south of Sutton Harbour, QAB promises a welcoming stay with alongside pontoon berthing protected by the surrounding breakwater. While on the city's doorstep, the marina is a short walk away from the historical Barbican, providing peace and tranquility to visitors.

The marina is often frequented by crowds of people marveling the many prestigious international yacht and powerboat races Plymouth Sound facilitates.

QAB is an exposed treasure along a beautiful historic coastline, at times resembling a mini 'Cowes' with its vibrancy.

FACILITIES AT A GLANCE

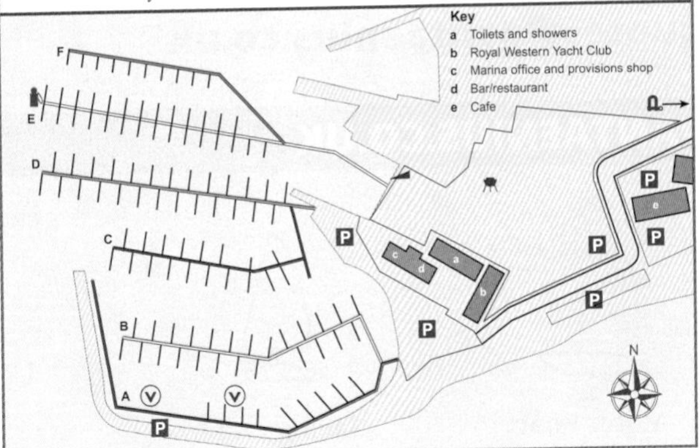
Key
a Toilets and showers
b Royal Western Yacht Club
c Marina office and provisions shop
d Bar/restaurant
e Cafe
N

PLYMOUTH YACHT HAVEN

Plymouth Yacht Haven Ltd
Shaw Way, Mount Batten, Plymouth, PL9 9XH
Tel: 01752 404231 www.yachthavens.com
Email: enquiries@plymouthyachthaven.com

VHF	Ch 80
ACCESS	H24

Situated minutes from Plymouth Sound, Plymouth Yacht Haven enjoys a tranquil setting, yet is just a five minute water taxi ride from the bustling Barbican with all its restaurants and attractions. The Yacht Haven offers excellent protection from the prevailing SW winds and is within easy reach of some of the most fantastic cruising grounds.

This 450-berth marina can accommodate vessels up to 45m in length and 7m in draught. Members of staff are on site 24/7 to welcome you as a visitor and to serve diesel. With an on site 75T travel hoist and storage, a restaurant, chandlery, and extensive range of marine service, Plymouth Yacht Haven has plenty to offer both on and off the water.

FACILITIES AT A GLANCE

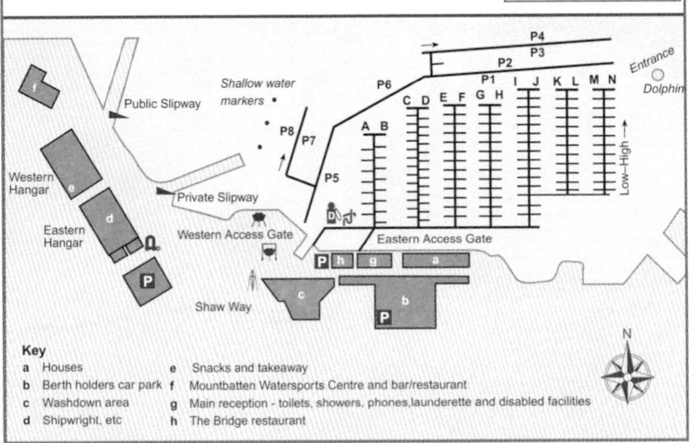
Public Slipway
Shallow water markers
Western Hangar
Eastern Hangar
Private Slipway
Western Access Gate
Eastern Access Gate
Shaw Way
Entrance
Dolphin
Low-High
P1
P2
P3
P4
P5
P6
P7
P8
A B
C D E F G H
I J K L M N
N
Key
a Houses
b Berth holders car park
c Washdown area
d Shipwright, etc
e Snacks and takeaway
f Mountbatten Watersports Centre and bar/restaurant
g Main reception - toilets, showers, phones,launderette and disabled facilities
h The Bridge restaurant

SUTTON HARBOUR

Sutton Harbour
The Jetty, Sutton Harbour, Plymouth, PL4 0DW
Tel: 01752 204702 Fax: 01752 204693
Email: marina@sutton-harbour.co.uk
www.suttonharbourmarina.com

VHF	Ch 12
ACCESS	H24

Sutton Harbour Marina, located in the heart of Plymouth's historic Barbican area and a short stroll from the city centre, offers 5-star facilities in a sheltered location, surrounded by boutique waterfront bars and restau
a minimum 3.5m depth, the harbour has 24hr lock access on request with free flow approx 3hrs either side of high tide. The marina of choice for international yacht races such as The Transat and Fastnet. Visitors are invited to come and enjoy the unrivalled shelter, facilities, atmosphere and location that Sutton harbour offers.

FACILITIES AT A GLANCE

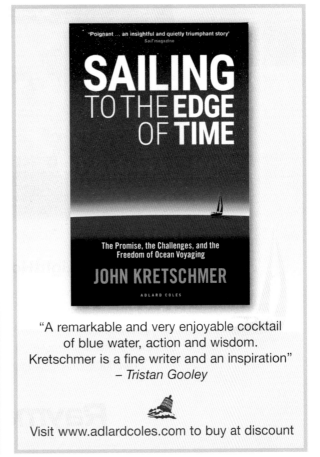

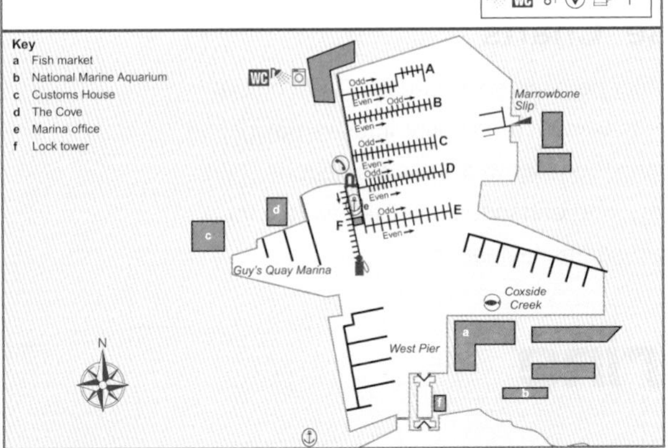
Key
a Fish market
b National Marine Aquarium
c Customs House
d The Cove
e Marina office
f Lock tower
Marrowbone Slip
Guy's Quay Marina
Coxside Creek
West Pier
A
B
C
D
E
F
Odd
Even
N

DARTHAVEN MARINA

Darthaven Marina
Brixham Road, Kingswear, Devon, TQ6 0SG
Tel: 01803 752242
Email: admin@darthaven.co.uk
www.darthaven.co.uk

VHF	Ch 80
ACCESS	H24

Darthaven Marina is a family run business situated in the village of Kingswear on the east side of the River Dart. Within half a mile from Start Bay and the mouth of the river, it is the first marina you come to from seaward and is accessible at all states of the tide. Darthaven prides itself on being more than just a marina, offering a high standard of marine services with both electronic and engineering experts plus wood and GRP repairs on site. A shop, post office and three pubs are within a walking distance of the marina, while a frequent ferry service takes passengers across the river to Dartmouth.

FACILITIES AT A GLANCE

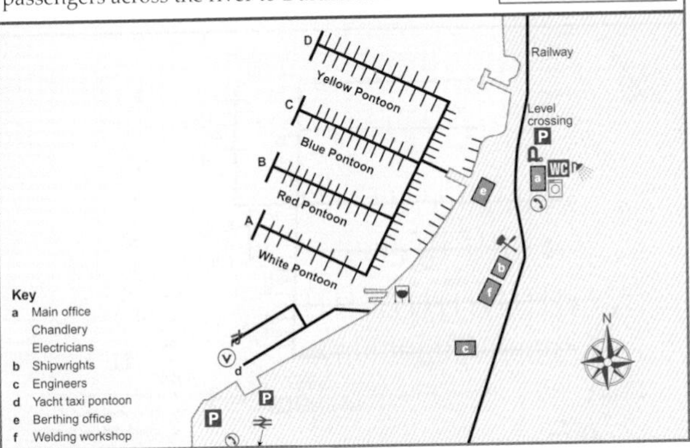
Railway
Level crossing
Yellow Pontoon
Blue Pontoon
Red Pontoon
White Pontoon
D
C
B
A
N
Key
a Main office
Chandlery
Electricians
b Shipwrights
c Engineers
d Yacht taxi pontoon
e Berthing office
f Welding workshop

DART MARINA

Dart Marina
Sandquay Road, Dartmouth, Devon, TQ6 9PH
Tel: 01803 837161 Fax: 01803 835040
Email: yachtharbour@dartmarina.com
www.dartmarinayachtharbour.com

VHF	Ch 80
ACCESS	H24

Dart Marina Yacht Harbour, in one of the most stunning locations on the UK coastline, is a peaceful spot for simply sitting on deck relaxing and perfectly positioned for day sailing or more ambitious cruising. With visitor berths, 110 annual berths, all accessible at any tide, the Yacht Harbour is sought after for its intimate atmosphere and stylish setting.

Professional, knowledgeable and helpful, the marina team is on-site all year round and there are impeccable facilities including showers, bathrooms and laundry.

In Dartmouth there are restaurants, bistros, cafes, delis, independent shops, galleries, a cinema, chandlers, antique and lifestyle shops in abundance.

transeurope

FACILITIES AT A GLANCE

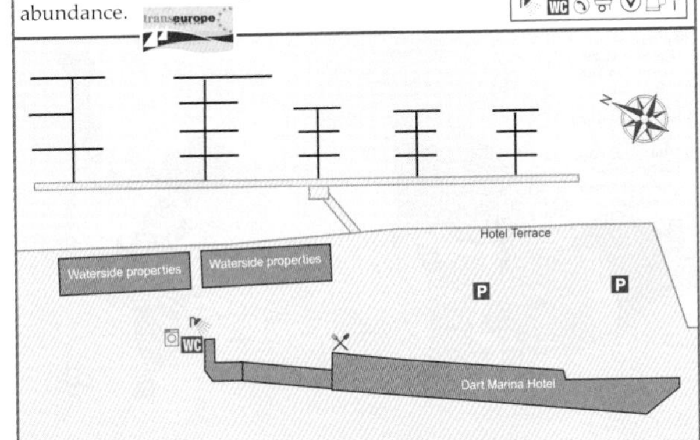
Hotel Terrace
Waterside properties
Waterside properties
Dart Marina Hotel

NOSS ON DART MARINA

Noss on Dart Marina
Bridge Road, Kingswear, Devon, TQ6 0EA
Tel: 01803 839087
Email: noss@premiermarinas.com
www.premiermarinas.com

VHF	Ch 80
ACCESS	H24

Set on the eastern bank of the River Dart, in a secluded area of outstanding natural beauty, Noss on Dart Marina offers easy access to some of the most beautiful anchorages and ports in Devon. Located close to woodland walks along the Dart Valley Trail, the area is a hotspot for local festivals and regattas.

A £75m development of the marina is underway with the 232-berth floating marina and fully-serviced boatyard set to be completed by summer 2021.

FACILITIES AT A GLANCE

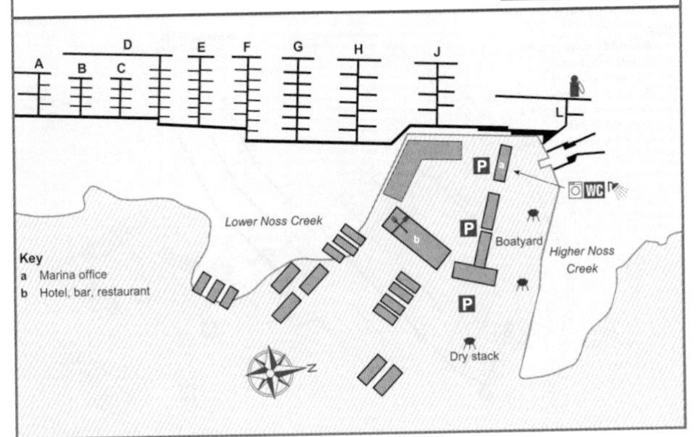
A
B
C
D
E
F
G
H
J
Lower Noss Creek
Boatyard
Higher Noss Creek
Dry stack
Key
a Marina office
b Hotel, bar, restaurant

DARTSIDE QUAY

Dartside Quay
Galmpton Creek, Brixham, Devon, TQ5 0EH
Tel: 01803 845445
Email: dartsidequay@mdlmarinas.co.uk
www.dartsidequay.co.uk

VHF	Ch 80
ACCESS	HW±2

Located at the head of Galmpton Creek, Dartside Quay lies three miles upriver from Dartmouth. In a sheltered position and with beautiful views across to Dittisham, it offers extensive boatyard facilities. The seven-acre dry boat storage area has space for over 300 boats and is serviced by a 65-ton hoist operating from a purpose-built dock, plus a 20-ton trailer hoist operating on a slipway. There are also a number of overnight/monthly pontoons available (dries out) in the summer months. You will also find various specialist marine tenants on site able to assist with boat repairs and maintenance.

FACILITIES AT A GLANCE

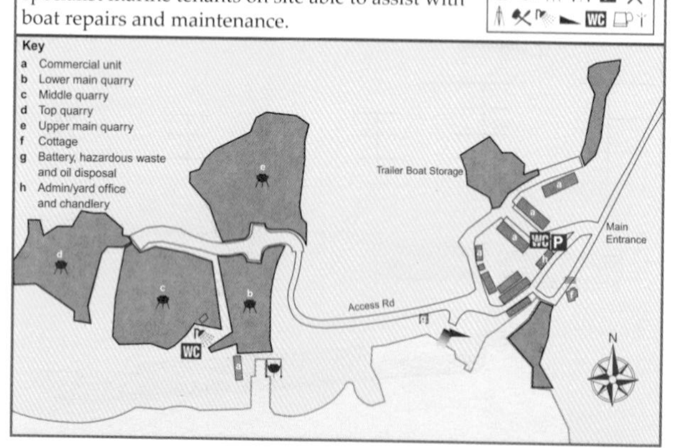
Key
a Commercial unit
b Lower main quarry
c Middle quarry
d Top quarry
e Upper main quarry
f Cottage
g Battery, hazardous waste and oil disposal
h Admin/yard office and chandlery
Trailer Boat Storage
Main Entrance
Access Rd
WC
P
N

BRIXHAM MARINA

Brixham Marina
Berry Head Road, Brixham, Devon, TQ5 9BW
Tel: 01803 882929
Email: brixham@mdlmarinas.co.uk
www.brixhammarina.co.uk

VHF	Ch 80
ACCESS	H24

Home to one of Britain's largest fishing fleets, Brixham Harbour is located on the southern shore of Torbay, which is well sheltered from westerly winds and where tidal streams are weak. Brixham Marina, housed in a separate basin to the work boats, provides easy access in all weather conditions and at all states of the tide. Provisions and diesel are available and there is a bar and restaurant, ideal for when you've worked up an appetite out on the water. Local attractions include Berry Head Nature Reserve and a visit to the replica of Francis Drake's *Golden Hind*.

FACILITIES AT A GLANCE

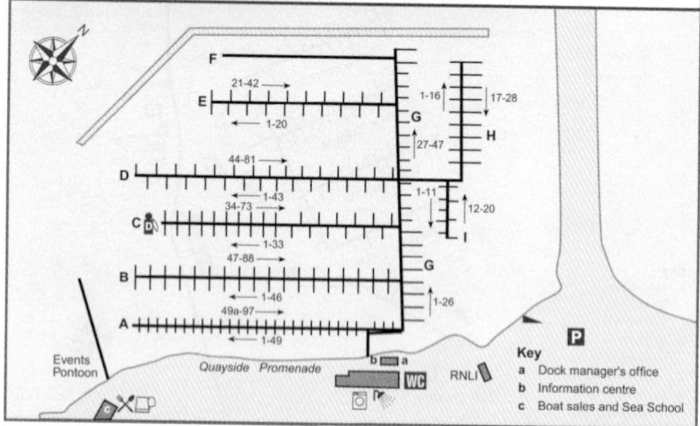
N
F
E
D
C
B
A
G
H
I
21-42
1-20
44-81
1-43
34-73
1-33
47-88
1-46
49a-97
1-49
1-16
17-28
27-47
1-11
12-20
1-26
Events Pontoon
Quayside Promenade
RNLI
WC
P
Key
a Dock manager's office
b Information centre
c Boat sales and Sea School

TORQUAY MARINA

Torquay Marina
Torquay, Devon, TQ2 5EQ
Tel: 01803 200210
Email: torquaymarina@mdlmarinas.co.uk
www.torquaymarina.co.uk

VHF	Ch 80
ACCESS	H24

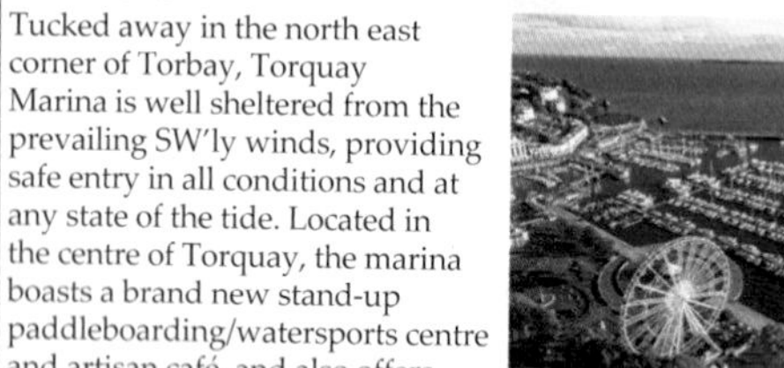

Tucked away in the north east corner of Torbay, Torquay Marina is well sheltered from the prevailing SW'ly winds, providing safe entry in all conditions and at any state of the tide. Located in the centre of Torquay, the marina boasts a brand new stand-up paddleboarding/watersports centre and artisan café, and also offers easy access to the town's numerous shops, bars and restaurants.

Torquay is ideally situated for either exploring Tor Bay itself, with its many delightful anchorages, or else for heading west to experience several other scenic harbours such as Dartmouth and Salcombe. It also provides a good starting point for crossing to Brittany, Normandy or the Channel Islands.

FACILITIES AT A GLANCE

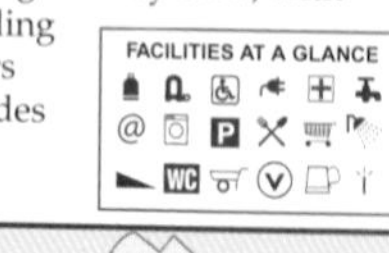

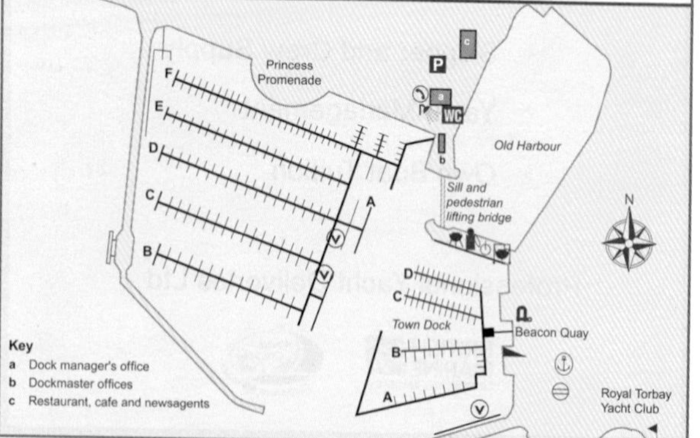
Princess Promenade
Old Harbour
Sill and pedestrian lifting bridge
Town Dock
Beacon Quay
Royal Torbay Yacht Club
N
F
E
D
C
B
A
Key
a Dock manager's office
b Dockmaster offices
c Restaurant, cafe and newsagents

PORTLAND MARINA

Portland Marina
Osprey Quay, Portland, Dorset, DT5 1DX
Tel: 01305 866190
Email: portland@boatfolk.co.uk
www.boatfolk.co.uk/portlandmarina

VHF	Ch 80
ACCESS	H24

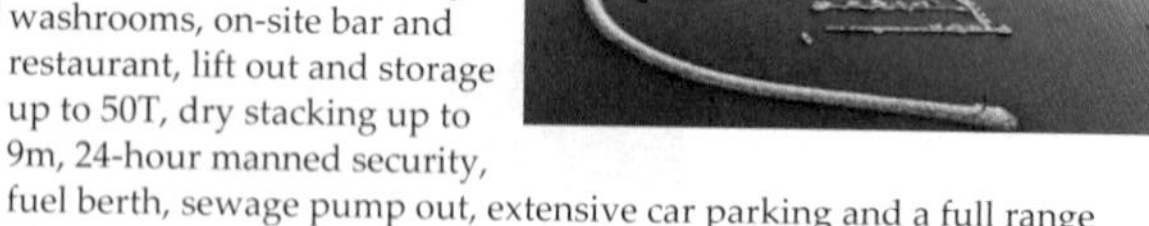

Portland Marina is an ideal location for both annual berthing and weekend stopovers. The marina offers first class facilities including washrooms, on-site bar and restaurant, lift out and storage up to 50T, dry stacking up to 9m, 24-hour manned security, fuel berth, sewage pump out, extensive car parking and a full range of marine services including a chandlery.

The marina is within walking distance of local pubs and restaurants on Portland with Weymouth's bustling town centre and mainline railway station just a short bus or ferry ride away.

FACILITIES AT A GLANCE

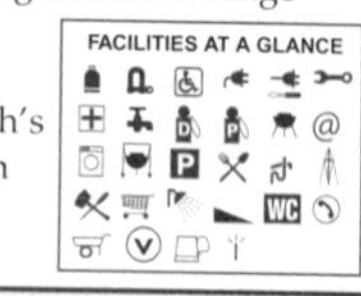

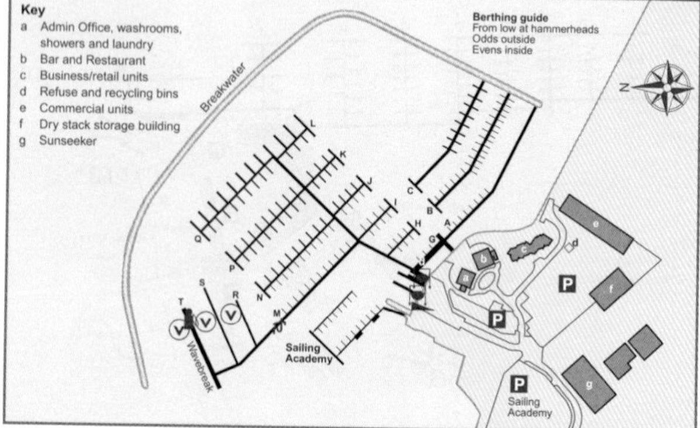
Key
a Admin Office, washrooms, showers and laundry
b Bar and Restaurant
c Business/retail units
d Refuse and recycling bins
e Commercial units
f Dry stack storage building
g Sunseeker
Berthing guide
From low at hammerheads
Odds outside
Evens inside
Breakwater
Wavebreak
Sailing Academy
N
P

WEYMOUTH HARBOUR

Harbour Office
13 Custom House Quay, Weymouth, Dorset, DT4 8BG
Tel: 01305 838386
Email: weymouthharbour@dorset.gov.uk
www.weymouth-harbour.co.uk

VHF Ch 12
ACCESS H24

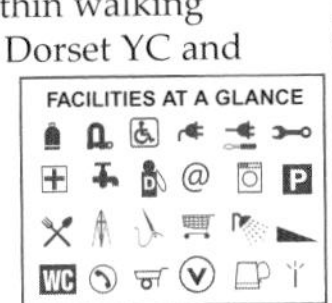

Weymouth Harbour, situated on the Jurassic Coast, lays NE of Portland in the protected waters of Weymouth Bay. Located in the heart of the old town and accessible at any state of tide, the Georgian harbour has numerous overnight berths. Pontoons on both quays have electricity and fresh water, as well as modern shower facilities. Restaurants and shops abound within walking distance and visitors are also welcome in the Royal Dorset YC and Weymouth SC, both situated on the quayside.

Vessels are advised to call *Weymouth Harbour* on Ch12 on approach, and vessels over 15m are recommended to give prior notification of intended arrival.

FACILITIES AT A GLANCE

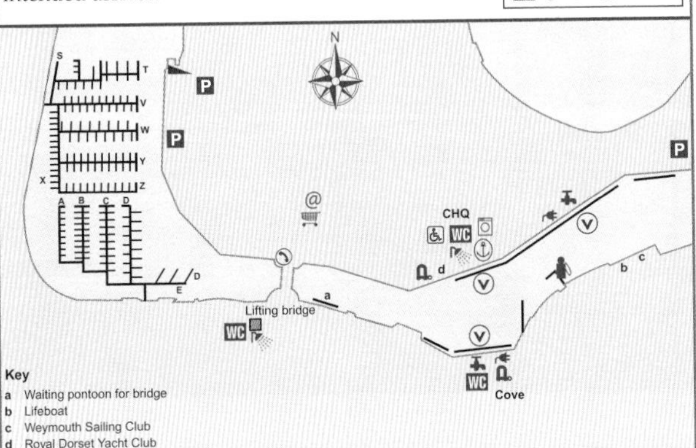
N
S
T
V
W
X
Y
Z
A
B
C
D
E
CHQ
Lifting bridge
Cove
Key
a Waiting pontoon for bridge
b Lifeboat
c Weymouth Sailing Club
d Royal Dorset Yacht Club

WEYMOUTH MARINA

Weymouth Marina
70 Commercial Road, Dorset, DT4 8NA
Tel: 01305 767576 Fax: 01305 767575
Email: weymouth@boatfolk.co.uk
www.boatfolk.co.uk/weymouthmarina

VHF Ch 80
ACCESS H24

With more than 280 permanent and visitors' berths, Weymouth is a modern, purpose-built marina ideally situated for yachtsmen cruising between the West Country and the Solent. It is also conveniently placed for sailing to France or the Channel Islands. Accessed via the town's historic lifting bridge, which opens every even hour 0800–2000 (plus 2100 Jun–Aug), the marina is dredged to 2.5m below chart datum. It provides easy access to the town centre, with its abundance of shops, pubs and restaurants, as well as to the traditional seafront where an impressive sandy beach is overlooked by an esplanade of hotels.

FACILITIES AT A GLANCE

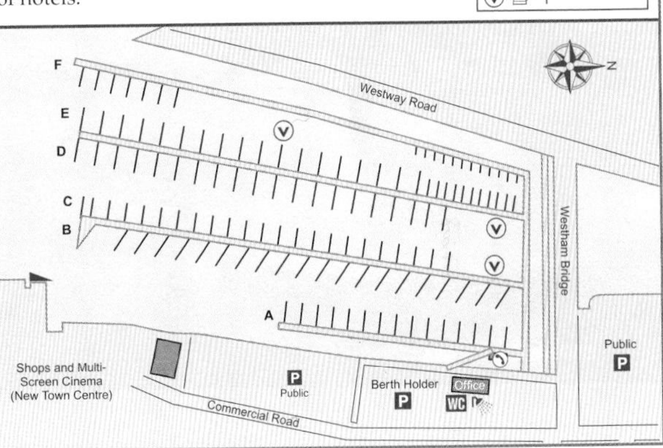
F
E
D
C
B
A
Westway Road
Westham Bridge
Shops and Multi-Screen Cinema (New Town Centre)
Public
Berth Holder
Office
Commercial Road
Public
Z

CENTRAL SOUTHERN ENGLAND – Anvil Point to Selsey Bill

Key to Marina Plans symbols

- Bottled gas
- Chandler
- Disabled facilities
- Electrical supply
- Electrical repairs
- Engine repairs
- First Aid
- Fresh Water
- Fuel - Diesel
- Fuel - Petrol
- Hardstanding/boatyard
- Internet Café
- Laundry facilities
- Lift-out facilities
- Parking
- Pub/Restaurant
- Pump out
- Rigging service
- Sail repairs
- Shipwright
- Shop/Supermarket
- Showers
- Slipway
- Toilets
- Telephone
- Trolleys
- Visitors berths
- Wi-Fi

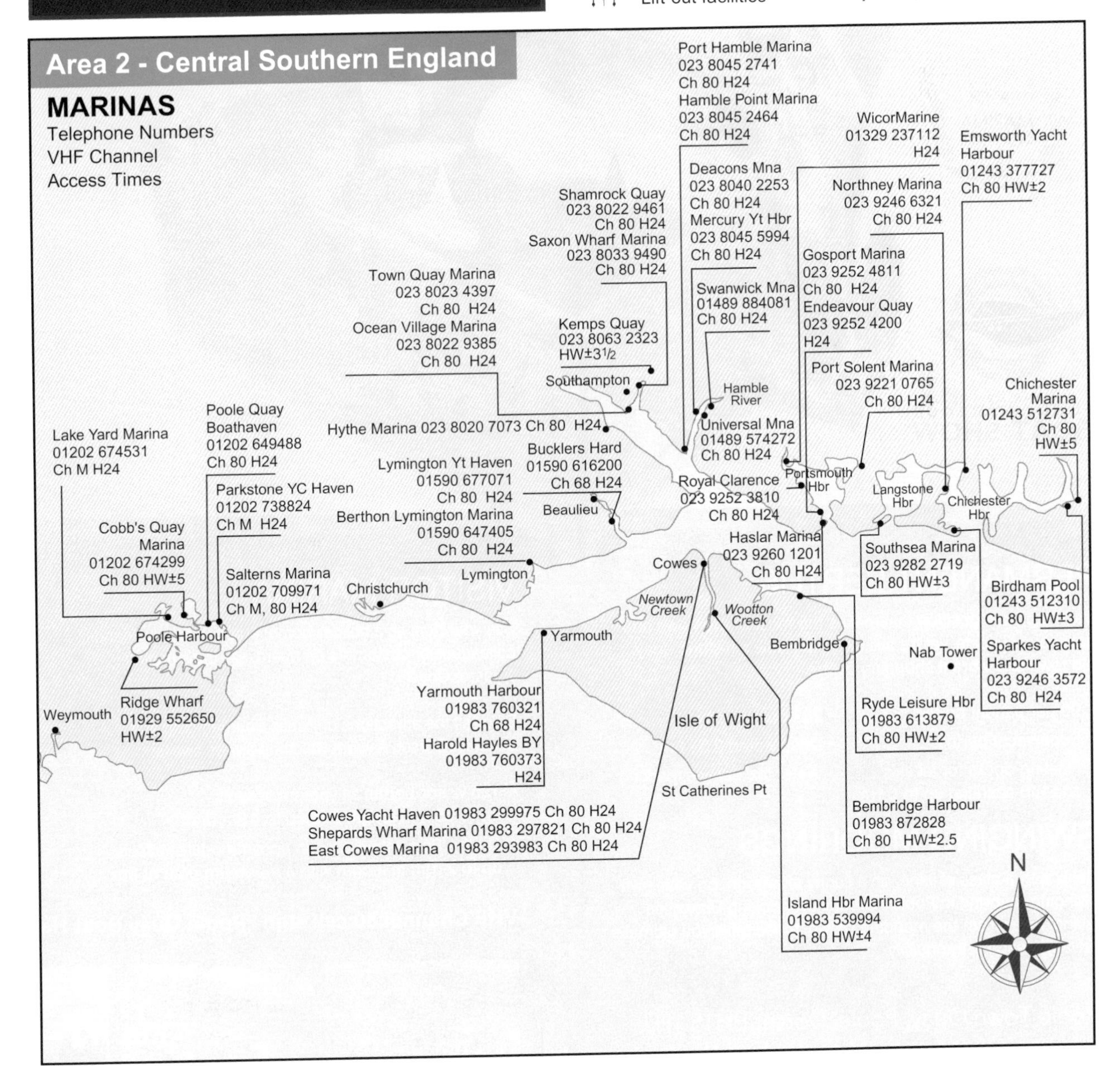
Area 2 - Central Southern England
MARINAS
Telephone Numbers
VHF Channel
Access Times
Port Hamble Marina 023 8045 2741 Ch 80 H24
Hamble Point Marina 023 8045 2464 Ch 80 H24
WicorMarine 01329 237112 H24
Emsworth Yacht Harbour 01243 377727 Ch 80 HW±2
Deacons Mna 023 8040 2253 Ch 80 H24
Mercury Yt Hbr 023 8045 5994 Ch 80 H24
Northney Marina 023 9246 6321 Ch 80 H24
Shamrock Quay 023 8022 9461 Ch 80 H24
Saxon Wharf Marina 023 8033 9490 Ch 80 H24
Gosport Marina 023 9252 4811 Ch 80 H24
Endeavour Quay 023 9252 4200 H24
Town Quay Marina 023 8023 4397 Ch 80 H24
Ocean Village Marina 023 8022 9385 Ch 80 H24
Swanwick Mna 01489 884081 Ch 80 H24
Kemps Quay 023 8063 2323 HW±3½
Port Solent Marina 023 9221 0765 Ch 80 H24
Southampton
Hamble River
Chichester Marina 01243 512731 Ch 80 HW±5
Poole Quay Boathaven 01202 649488 Ch 80 H24
Lake Yard Marina 01202 674531 Ch M H24
Hythe Marina 023 8020 7073 Ch 80 H24
Universal Mna 01489 574272 Ch 80 H24
Bucklers Hard 01590 616200 Ch 68 H24
Lymington Yt Haven 01590 677071 Ch 80 H24
Portsmouth Hbr
Royal Clarence 023 9252 3810 Ch 80 H24
Langstone Hbr
Chichester Hbr
Parkstone YC Haven 01202 738824 Ch M H24
Beaulieu
Berthon Lymington Marina 01590 647405 Ch 80 H24
Haslar Marina 023 9260 1201 Ch 80 H24
Southsea Marina 023 9282 2719 Ch 80 HW±3
Cobb's Quay Marina 01202 674299 Ch 80 HW±5
Cowes
Salterns Marina 01202 709971 Ch M, 80 H24
Lymington
Christchurch
Birdham Pool 01243 512310 Ch 80 HW±3
Newtown Creek
Wootton Creek
Poole Harbour
Yarmouth
Bembridge
Nab Tower
Sparkes Yacht Harbour 023 9246 3572 Ch 80 H24
Ridge Wharf 01929 552650 HW±2
Weymouth
Yarmouth Harbour 01983 760321 Ch 68 H24
Harold Hayles BY 01983 760373 H24
Isle of Wight
Ryde Leisure Hbr 01983 613879 Ch 80 HW±2
St Catherines Pt
Bembridge Harbour 01983 872828 Ch 80 HW±2.5
Cowes Yacht Haven 01983 299975 Ch 80 H24
Shepards Wharf Marina 01983 297821 Ch 80 H24
East Cowes Marina 01983 293983 Ch 80 H24
Island Hbr Marina 01983 539994 Ch 80 HW±4
N

RIDGE WHARF YACHT CENTRE

Ridge Wharf Yacht Centre
Ridge, Wareham, Dorset, BH20 5BG
Tel: 01929 552650 Fax: 01929 554434
Email: office@ridgewharf.co.uk www.ridgewharf.co.uk

VHF	
ACCESS	HW±2

On the south bank of the River Frome, which acts as the boundary to the North of the Isle of Purbeck, is Ridge Wharf Yacht Centre. Access for a 1.5m draught is between one and two hours either side of HW, with berths drying out to soft mud. The Yacht Centre cannot be contacted on VHF, so it is best to phone up ahead of time to inquire about berthing availability.

A trip upstream to the ancient market town of Wareham is well worth while, although owners of deep-draughted yachts may prefer to go by dinghy. Tucked between the Rivers Frome and Trent, it is packed full of cafés, restaurants and shops.

FACILITIES AT A GLANCE

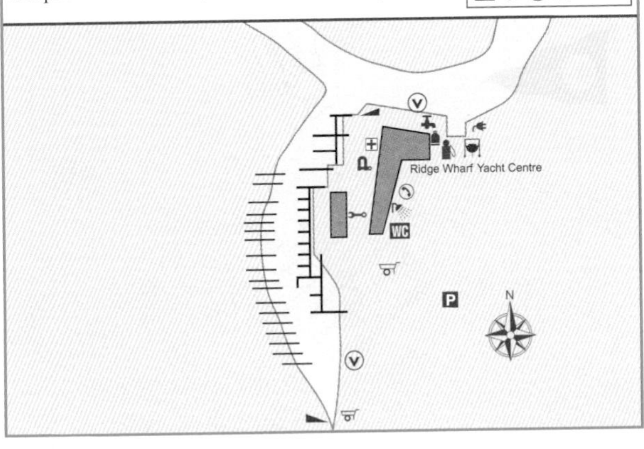
Ridge Wharf Yacht Centre
WC
P
N

LAKE YARD MARINA

Lake Yard Marina
Lake Drive, Hamworthy, Poole, Dorset BH15 4DT
Tel: 01202 674531
Email: office@lakeyard.com www.lakeyard.com

VHF	Ch M
ACCESS	H24

Lake Yard is situated towards the NW end of Poole Harbour, just beyond the SHM No 73. The entrance can be easily identified by 2FR (vert) and 2FG (vert) lights. Enjoying 24 hour access, the marina has no designated visitors' berths, but will accommodate visiting yachtsmen if resident berth holders are away. Its on site facilities include full maintenance and repair services as well as hard standing and a 50 ton boat hoist, although for the nearest fuel go to Corralls (Tel 01202 674551), opposite the Town Quay. Lake Yard's Waterfront Club, offering spectacular views across the harbour, opens 7/7 for breakfast, lunch and evening meals.

FACILITIES AT A GLANCE

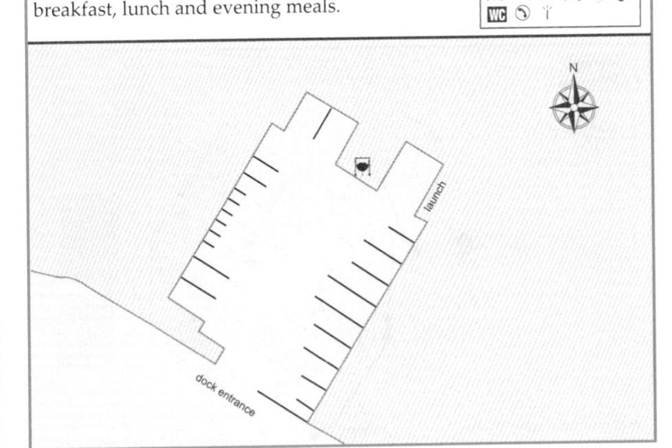
N
launch
dock entrance

COBB'S QUAY MARINA

Cobb's Quay Marina
Hamworthy, Poole, Dorset, BH15 4EL
Tel: 01202 674299
Email: cobbsquay@mdlmarinas.co.uk
www.cobbsquaymarina.co.uk

VHF	Ch 80
ACCESS	H24

Lying on the west side of Holes Bay in Poole Harbour, Cobb's Quay is accessed via two lifting bridges. The hourly lifting schedule which runs from 0530 to 2330 (except for weekday rush hours) makes the entrance into Holes Bay accessible 24/7. With fully serviced pontoons for yachts up to 20m LOA, visitors can enjoy the facilities including the highly reputable Cobb's YC. The marina also offers a convenient 240-berth dry stack area for motorboats up to 10m. With increased security and lower maintenance costs, the service includes unlimited launching on arrival. There is also a 40-ton hoist, fully serviced boatyard and on-site self-storage. Poole Harbour is the second largest natural harbour in the world and is rich in wildlife, water sports and secret hideaways.

FACILITIES AT A GLANCE

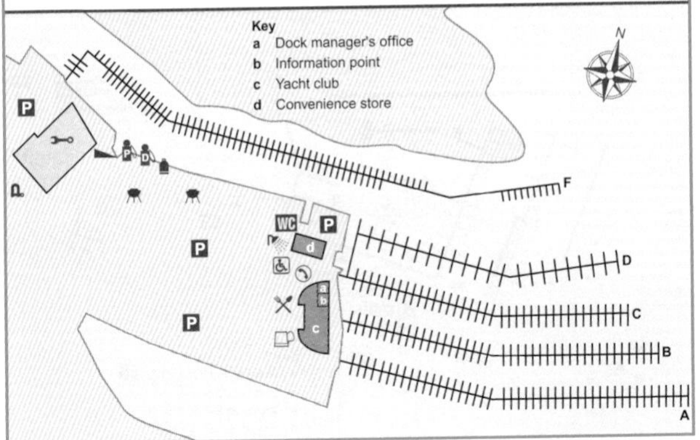
Key
a Dock manager's office
b Information point
c Yacht club
d Convenience store
N
P
WC
A
B
C
D
F

POOLE QUAY BOAT HAVEN

Poole Quay Boat Haven
Poole Town Quay, Poole, Dorset, BH15 1HJ
Tel: 01202 649488
Email: info@poolequayboathaven.co.uk

VHF	Ch 80
ACCESS	H24

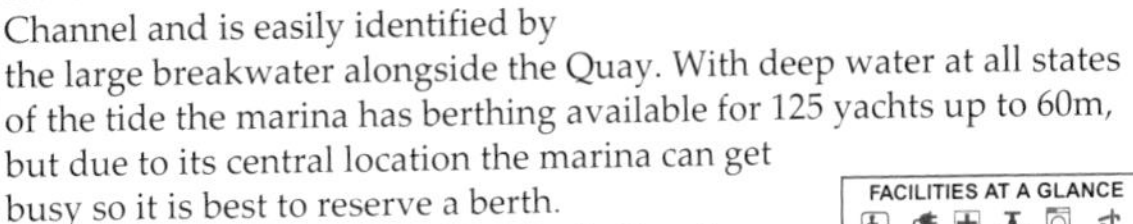

Once inside the Poole Harbour entrance small yachts heading for Poole Quay Boat Haven should use the Boat Channel running parallel south of the dredged Middle Ship Channel, which is primarily used by ferries sailing to and from the Hamworthy terminal. The marina can be accessed via the Little Channel and is easily identified by the large breakwater alongside the Quay. With deep water at all states of the tide the marina has berthing available for 125 yachts up to 60m, but due to its central location the marina can get busy so it is best to reserve a berth.

There is easy access to all of Poole Quay's facilities including restaurants, bars, Poole Pottery and the Waterfront Museum.

FACILITIES AT A GLANCE

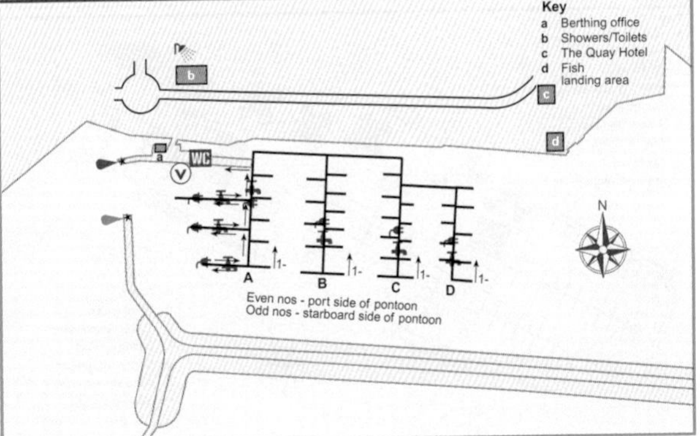
Key
a Berthing office
b Showers/Toilets
c The Quay Hotel
d Fish landing area
N
A
B
C
D
Even nos - port side of pontoon
Odd nos - starboard side of pontoon

PORT OF POOLE MARINA

Port of Poole Marina
Poole Town Quay, Poole, Dorset, BH15 1HJ
Tel: 01202 649488
Email: info@poolequayboathaven.co.uk

VHF Ch 80
ACCESS H24

Beware of the chain ferry operating at the entrance to Poole harbour. Once inside small yachts heading for the marina should use the Boat Channel running parallel south of the Middle Ship Channel. The marina is to the east of the main ferry terminals and can be identified by a large floating breakwater at the entrance.

The marina has deep water at all tides and berthing for 60 permanent vessels. It is also used as an overflow for visitors from Poole Quay Boat Haven, subject to availability.

A water taxi is available during daylight hours to access the quay for restaurants and shops, also accessible with a 10–15min walk round the quays.

FACILITIES AT A GLANCE

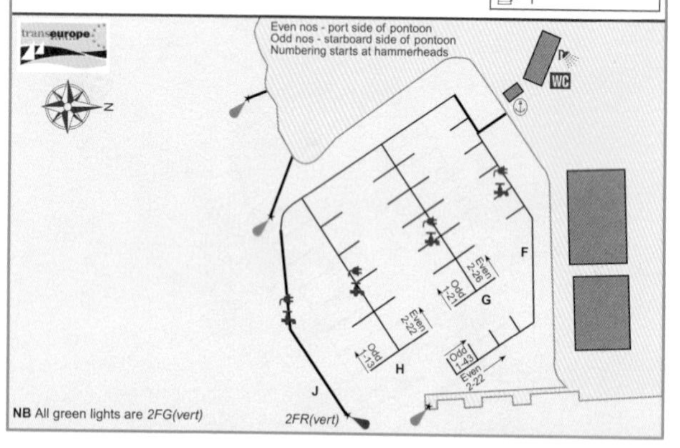
transeurope
Even nos - port side of pontoon
Odd nos - starboard side of pontoon
Numbering starts at hammerheads
WC
F
G
H
J
Even 2-26
Odd 1-21
Even 2-22
Odd 1-13
Odd 1-43
Even 2-22
NB All green lights are 2FG(vert)
2FR(vert)

PARKSTONE YACHT HAVEN

Parkstone Yacht Club
Pearce Avenue, Parkstone, Poole, Dorset, BH14 8EH
Tel: 01202 738824 Fax: 01202 716394
Email: office@parkstoneyc.co.uk

VHF	Ch M
ACCESS	H24

Situated on the north side of Poole Harbour between Salterns Marina and Poole Quay Boat Haven, Parkstone Yacht Haven can be entered at all states of the tides. Its approach channel has been dredged to 2.0m and is clearly marked by buoys. Run by the Parkstone Yacht Club, the Haven provides 200 deep water berths for members and visitors' berths. Other services include a new office facility with laundry and WCs, bar, restaurant, shower/changing rooms and wi-fi. With a busy sailing programme for over 2,500 members, the Yacht Club plays host to a variety of events including Poole Week, which is held towards the end of August. Please phone for availability.

FACILITIES AT A GLANCE

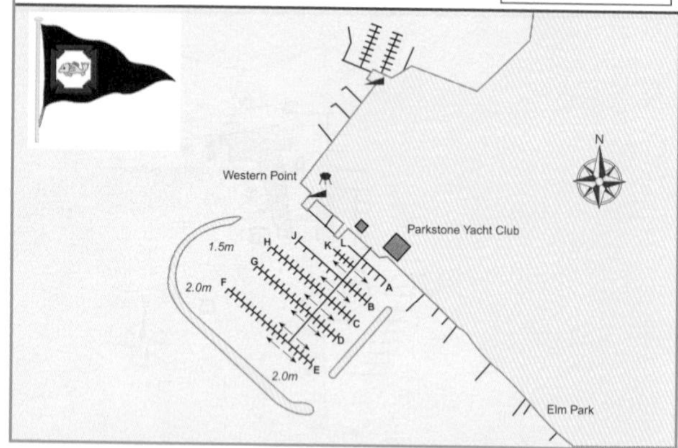
N
Western Point
Parkstone Yacht Club
1.5m
2.0m
2.0m
A
B
C
D
E
F
G
H
J
K
L
Elm Park

SALTERNS MARINA

Salterns Marina
40 Salterns Way, Lilliput, Poole
Dorset, BH14 8JR
Tel: 01202 709971 Fax: 01202 700398
Email: marina@salterns.co.uk www.salterns.co.uk

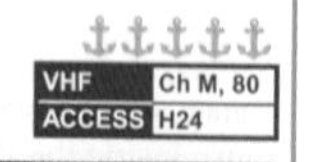
VHF Ch M, 80
ACCESS H24

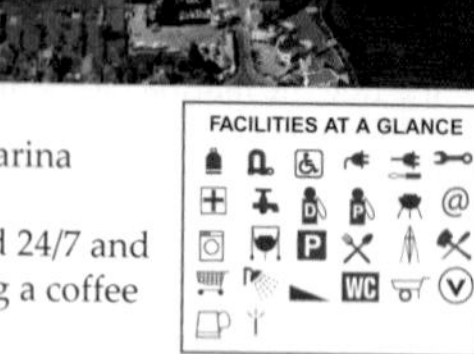

Holding the Five Gold Anchor award, Salterns Marina provides a service which is second to none. Located off the North Channel, it is approached from the No 31 SHM and benefits from deep water at all states of the tide. Facilities include 220 alongside pontoon berths as well as 75 swinging moorings with a free launch service. However, with very few designated visitors' berths, it is best to contact the marina ahead of time for availability.

Fuel, diesel and gas can all be obtained 24/7 and the well-stocked chandlery, incorporating a coffee shop, stays open seven days a week.

FACILITIES AT A GLANCE

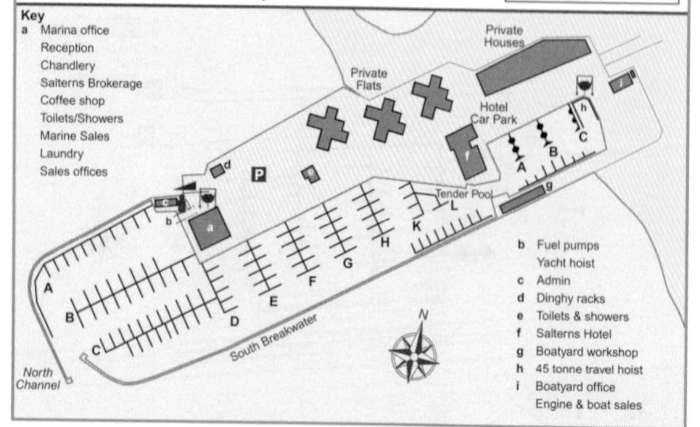
Key
a Marina office
Reception
Chandlery
Salterns Brokerage
Coffee shop
Toilets/Showers
Marine Sales
Laundry
Sales offices
b Fuel pumps
Yacht hoist
c Admin
d Dinghy racks
e Toilets & showers
f Salterns Hotel
g Boatyard workshop
h 45 tonne travel hoist
i Boatyard office
Engine & boat sales
Private Houses
Private Flats
Hotel Car Park
Tender Pool
South Breakwater
North Channel
N

YARMOUTH HARBOUR

Yarmouth Harbour
Yarmouth, Isle of Wight, PO41 0NT
Tel: 01983 760321
info@yarmouth-harbour.co.uk
www.yarmouth-harbour.co.uk

VHF Ch 68
ACCESS H24

The most western harbour on the Isle of Wight, Yarmouth is not only a convenient passage stopover but a very desirable destination in its own right, with virtually all weather and tidal access, although strong N to NE'ly winds can produce a considerable swell.

The HM launch patrols the harbour entrance and will direct visiting yachtsmen to a walkashore or standalone pontoon berths; call *Yarmouth Harbour* on Ch 68 prior to entering the harbour. The pretty harbour and town offer plenty of fine restaurants and amenities.

There are two boat yards in Yarmouth, River Yar boat yard in the SW corner of the harbour, and Harold Hayles Boat yard to the West.

FACILITIES AT A GLANCE

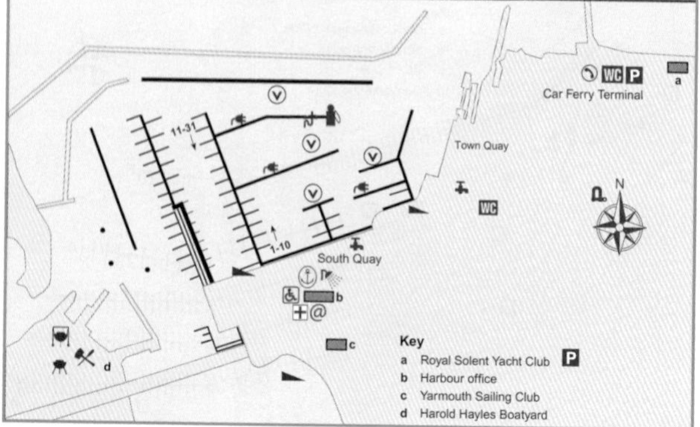
Car Ferry Terminal
Town Quay
South Quay
11-31
1-10
N
Key
a Royal Solent Yacht Club
b Harbour office
c Yarmouth Sailing Club
d Harold Hayles Boatyard

WELCOME
TO
BERTHON

LYMINGTON YACHT HAVEN

Lymington Yacht Haven
King's Saltern Road, Lymington, S041 3QD
Tel: 01590 677071
Email: lymington@yachthavens.com www.yachthavens.com

VHF Ch 80
ACCESS H24

Lymington Yacht Haven has the enviable position of being the first marina that comes into sight on your port hand side as you make your way up the well-marked Lymington river channel. Nestled between the 500 acre Lymington to Keyhaven nature reserve and the famous Georgian market town of Lymington, there is something for everyone.

Lymington Yacht Haven is manned 24/7 for fuel and berthing and boasts the most modern luxury shore side facilities you will find in the UK. Bike and electric bike hire are available through the marina office to explore the beautiful New Forest.

FACILITIES AT A GLANCE

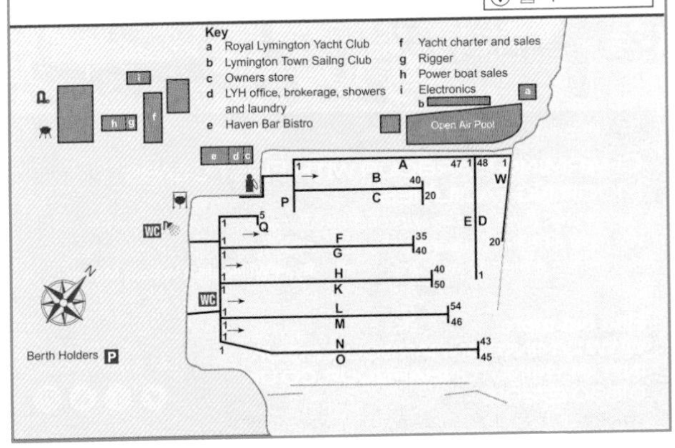
Key
a Royal Lymington Yacht Club
b Lymington Town Sailing Club
c Owners store
d LYH office, brokerage, showers and laundry
e Haven Bar Bistro
f Yacht charter and sales
g Rigger
h Power boat sales
i Electronics
Open Air Pool
Berth Holders
WC

BERTHON LYMINGTON MARINA

Berthon Lymington Marina Ltd
The Shipyard, Lymington, Hampshire, SO41 3YL
Tel: 01590 647405 Fax: 01590 647446
www.berthon.co.uk Email: marina@berthon.co.uk

VHF Ch 80
ACCESS H24

Situated approximately half a mile up river of Lymington Yacht Haven, on the port hand side, is Lymington Marina. Easily accessible at all states of the tide, it offers between 60 to 70 visitors' berths, with probably the best washrooms in the Solent. Its close proximity to the town centre and first rate services mean that booking is essential on busy weekends. Lymington Marina's parent, Berthon Boat Co, has state of the art facilities and a highly skilled work force of 100+ to deal with any repair, maintenance or refit.

Lymington benefits from having the New Forest on its doorstep and the Solent Way footpath provides an invigorating walk to and from Hurst Castle.

FACILITIES AT A GLANCE

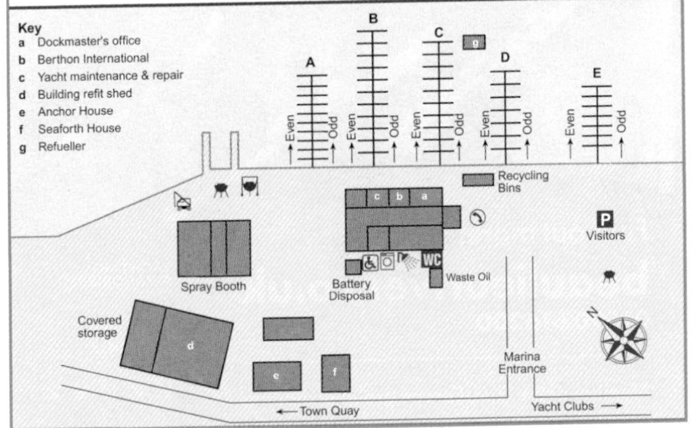
Key
a Dockmaster's office
b Berthon International
c Yacht maintenance & repair
d Building refit shed
e Anchor House
f Seaforth House
g Refueller
Even
Odd
Recycling Bins
Visitors
Waste Oil
Spray Booth
Battery Disposal
Covered storage
Marina Entrance
Town Quay
Yacht Clubs

LYMINGTON TOWN QUAY

Lymington Harbour Commissioners
Bath Road, Lymington, S041 3SE
Tel: 01590 672014
Email: info@lymingtonharbour.co.uk

VHF Ch 66
ACCESS H24

In 2020, the Town Quay pontoon was extended to provide 46 walk ashore berths for visiting boats, including 26 finger berths. All have power, water and free wi-fi. The Town Quay also has 32 fore and aft visitor moorings.

The Quay provides easy access to the historic cobbles and the attractive Georgian high street with its bars, restaurants and shops. The facilities of the town are close by with the historic Saturday market in the High Street. The New Forest is easily accessible by bus train and bicycle.

FACILITIES AT A GLANCE

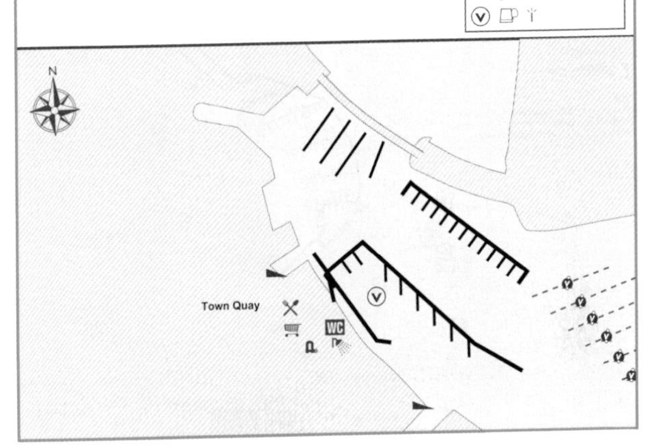
N
Town Quay

LYMINGTON – DAN BRAN VISITOR PONTOON

Lymington Harbour Commissioners
Bath Road, Lymington, S041 3SE
Tel: 01590 672014
Email: info@lymingtonharbour.co.uk

VHF	Ch 66
ACCESS	H24

Lymington Harbour Commission provides dedicated walk ashore visitor berths at the Lymington Town Quay and at the Dan Bran pontoon which comes ashore adjacent to the Royal Lymington YC and Lymington Town SC. Dan Bran is accessible at all states of the tide inside the wave screen. Power is available along its 650′ length with use of the facilities at Lymington Town SC and walk ashore access to the town and nearby nature reserves on the salt marsh. Sited between the marinas a short walk from Town Quay, it is ideal for club rallies/events and can accommodate up to 50 boats together. The sea water swimming bath adjacent to the pontoon is a popular venue for children and families.

FACILITIES AT A GLANCE
WC

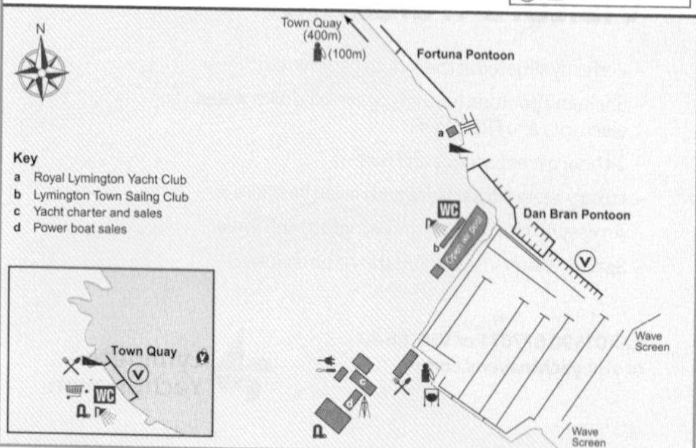
N
Town Quay (400m)
(100m)
Fortuna Pontoon
Key
a Royal Lymington Yacht Club
b Lymington Town Sailing Club
c Yacht charter and sales
d Power boat sales
WC
Dan Bran Pontoon
Town Quay
Wave Screen
Wave Screen

LYMINGTON HARBOUR
ONLY A STEP AWAY...
Beautiful Georgian Town
New Forest Activities
Fantastic Dining Experiences
Things To See & Do
NEW walk ashore berths at Town Quay with power, water & Wi-Fi
CHARMING HARBOUR WITH FIRST CLASS MOORING FACILITIES
"Lymington River and its approaches offer a safe haven for yachtsmen. We have a beautiful river and harbour area, and we pride ourselves on offering a friendly welcome to all." Ryan Willegers, Harbour Master
Find out more:
WWW.LYMINGTONHARBOUR.CO.UK
INFO@LYMINGTONHARBOUR.CO.UK
01590 672014
LET'S GET SOCIAL

BUCKLERS HARD MARINA

Bucklers Hard
Beaulieu, New Forest, Hampshire, SO42 7XB
Tel: 01590 616200
www.beaulieuriver.co.uk harbour.office@beaulieu.co.uk

VHF	Ch 68
ACCESS	H24

Situated on the Beaulieu River, Buckler's Hard Yacht Harbour is the ideal location from which to sail in the Solent or visit for a short stay to explore the surrounding New Forest. Recent investment in the TYHA 5 Gold Anchor marina offers improved facilities and technology to keep pace with modern demands, while preserving the unique character and charm of the unspoilt natural haven. Both permanent marina berths and river moorings are available. Visiting boats are welcome and advised to book in advance or radio Beaulieu River Radio on CH68 before entering the river

FACILITIES AT A GLANCE
WC

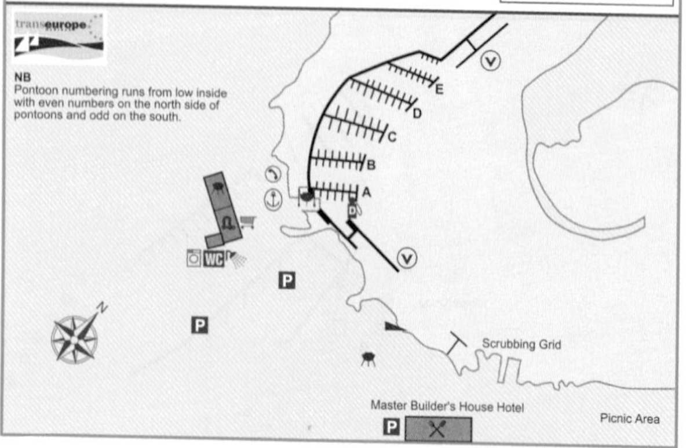
transeurope
NB
Pontoon numbering runs from low inside with even numbers on the north side of pontoons and odd on the south.
E
D
C
B
A
WC
P
P
Scrubbing Grid
Master Builder's House Hotel
Picnic Area

BEAULIEU RIVER
BUCKLER'S HARD YACHT HARBOUR
Celebrating 50 years 1971 - 2021
Find out more:
beaulieuriver.co.uk
01590 616200
THE BRITISH YACHTING AWARDS 2020 RUNNER-UP
GOLD ANCHOR
Member of the TransEurope group of marinas

HYTHE MARINA VILLAGE

Hythe Marina Village
Shamrock Way, Hythe, Southampton, SO45 6DY
Tel: 023 8020 7073
Email: hythe@mdlmarinas.co.uk
www.hythemarinavillage.co.uk

VHF	Ch 80
ACCESS	H24

Situated on the western shores of Southampton Water, Hythe Marina Village is approached by a dredged channel leading to a lock basin. The lock gates are controlled H24 throughout the year with a waiting pontoon south of the approach basin.

Hythe Marina Village incorporates full marine services as well as on-site restaurants and a selection of shops can be found in the town centre, a 5 minute walk away. Forming an integral part of the New Forest Waterside, Hythe is the perfect base from which to explore Hampshire's pretty inland villages and towns, or alternatively you can catch the ferry to Southampton's Town Quay.

FACILITIES AT A GLANCE
WC

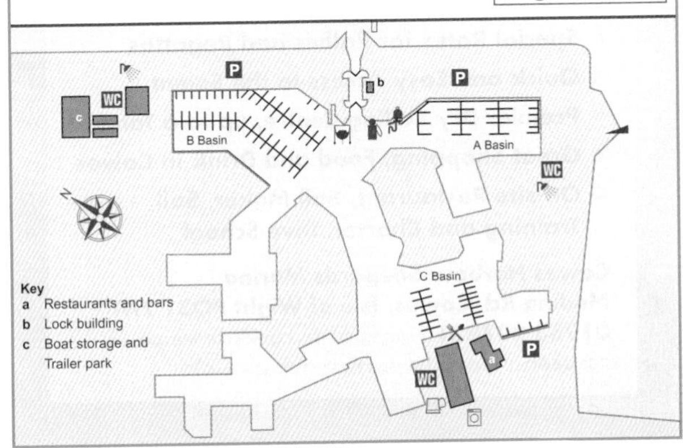
B Basin
A Basin
C Basin
Key
a Restaurants and bars
b Lock building
c Boat storage and Trailer park

COWES YACHT HAVEN

Cowes Yacht Haven
Vectis Yard, Cowes, Isle of Wight, PO31 7BD
Tel: 01983 299975 Fax: 01983 200332
www.cowesyachthaven.com
Email: info@cowesyachthaven.com

VHF	Ch 80
ACCESS	H24

Situated virtually at the centre of the Solent, Cowes is best known as Britain's premier yachting centre and offers all types of facilities to yachtsmen. Cowes Yacht Haven, operating 24 hours a day, has very few permanent moorings and is dedicated to catering for visitors and events. At peak times it can become very crowded and for occasions such as Aberdeen Asset Management Cowes Week you need to book up in advance.

FACILITIES AT A GLANCE
WC

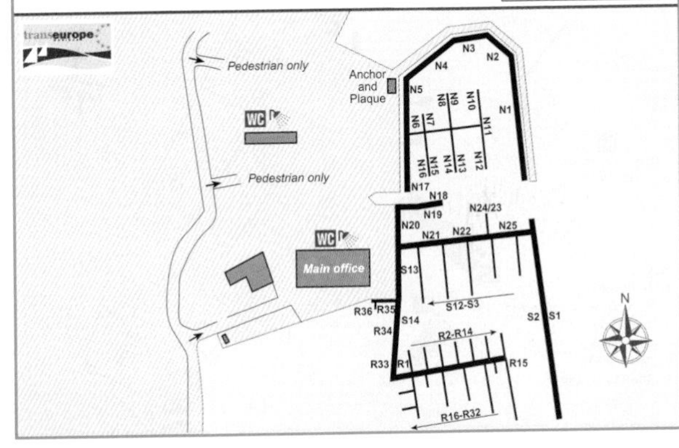
transeurope
Pedestrian only
Anchor and Plaque
Pedestrian only
Main office
S12-S3
R2-R14
R16-R32
N

ISLAND HARBOUR MARINA

Island Harbour Marina
Mill Lane, Binfield, Newport, Isle of Wight, PO30 2LA
Tel: 01983 539994 Fax: 01983 523401
Email: info@island-harbour.co.uk

VHF	Ch 80
ACCESS	HW±4

Situated in beautiful rolling farmland about half a mile south of Folly Inn, Island Harbour Marina provides around 200 visitors' berths. Protected by a lock that is operated daily from 0800 – 2100 during the summer and from 0800 – 1730 during the winter, the marina is accessible for about three hours either side of HW for draughts of 1.5m.

Due to its secluded setting, the marina's on site chandlery also sells essential provisions and newspapers. A half hour walk along the river brings you to Newport, the capital and county town of the Isle of Wight.

FACILITIES AT A GLANCE

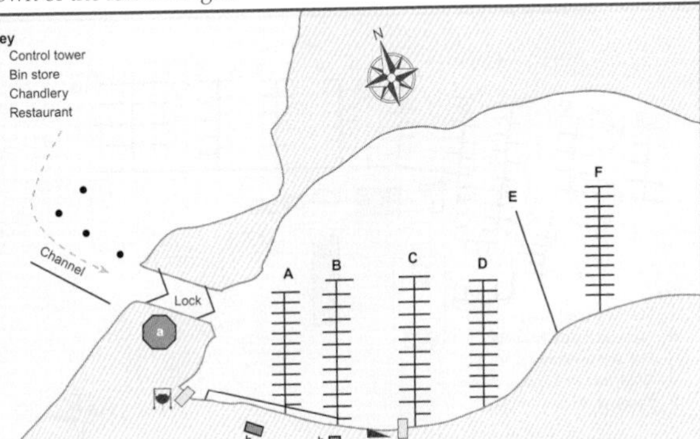
Key
a Control tower
b Bin store
c Chandlery
d Restaurant
N
Channel
Lock
A
B
C
D
E
F

COWES HARBOUR SHEPARDS MARINA

Cowes Harbour Shepards Marina
Medina Road, Cowes, Isle of Wight, PO31 7HT
Tel: 01983 297821 Email: shepards.chc@cowes.co.uk
www.cowesharbourshepardsmarina.co.uk

VHF	Ch 80
ACCESS	H24

Shepards Marina is one of Cowes Harbour's main marina facilities offering services and amenities for yacht racing events, rallies, and catering also to the cruising sailor and powerboater. The marina has capacity for 130 visiting boats, and 40 resident berth holders.

Visitor berths can be booked in advance, subject to availability. All berths benefit from water and electricity, free Wi-Fi, inclusive showers, and site-wide CCTV. Discounted rates are available for rallies of six or more boats, sailing schools, and winter berthing.

On site are The Basque Kitchen, Salty Sailing, Island Divers, and Solent Sails.

Fuel can be obtained from the Cowes Harbour Services Fuel Berth, 200m south of the Chain Ferry.

FACILITIES AT A GLANCE

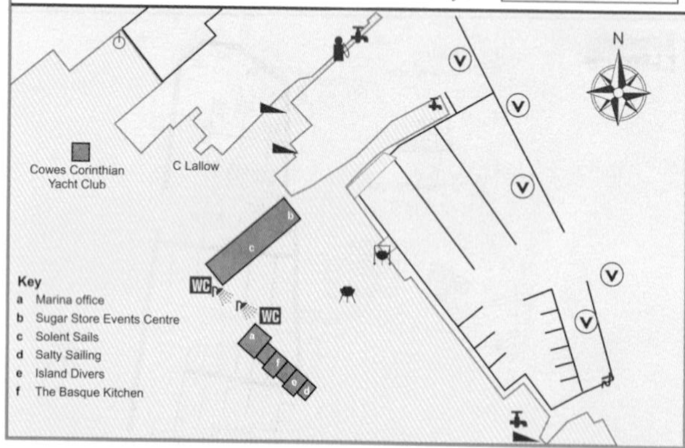
N
Cowes Corinthian Yacht Club
C Lallow
Key
a Marina office
b Sugar Store Events Centre
c Solent Sails
d Salty Sailing
e Island Divers
f The Basque Kitchen

COWES HARBOUR
SHEPARDS MARINA
DRY SAILING · EVENTS · MOORINGS
VISITORS · RALLIES · REGATTAS · WELCOME!
Top Staycation Destination
Pre-Book Overnight Visitor Berths
Special Rates for Rallies and Regattas
Quick and Easy Access to the Solent
Premier Dry Sailing Centre, up to 6 Tonnes
Great Shopping, Food and Drink in Cowes
On-site Restaurant, Sail Maker, Sail Training and Charter, Dive School
Cowes Harbour Shepards Marina
Medina Rd, Cowes, Isle of Wight PO31 7HT
01983 297821 | shepards.chc@cowes.co.uk
cowesharbourshepardsmarina.co.uk

EAST COWES MARINA

East Cowes Marina
Britannia Way, East Cowes, Isle of Wight, PO32 6UB
Tel: 01983 293983
Email: eastcowes@boatfolk.co.uk
www.boatfolk.co.uk/eastcowesmarina

VHF	Ch 80
ACCESS	H24

Accommodating around 235 residential yachts and 150 visiting boats at all states of the tide, East Cowes Marina is situated on the quiet and protected east bank of the Medina River, about a quarter mile above the chain ferry. A small convenience store is just five minutes walk away. The new centrally heated shower and toilet facilities ensure the visitor a warm welcome at any time of the year, as does the on-site pub and restaurant.

Several water taxis provide a return service to Cowes, ensuring a quick and easy way of getting to West Cowes.

FACILITIES AT A GLANCE

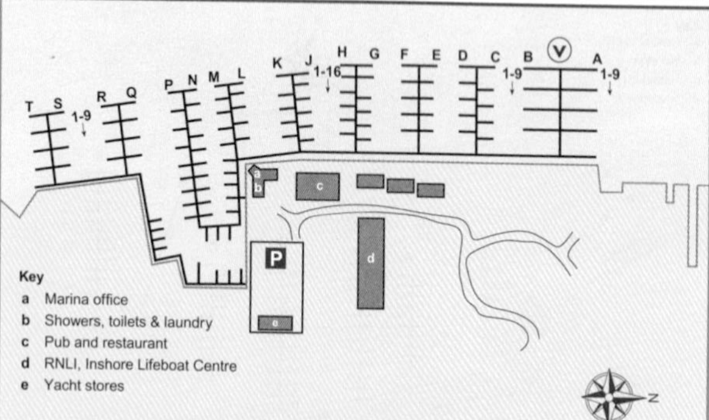
T S R Q P N M L K J H G F E D C B A
1-9
1-16
1-9
1-9
Key
a Marina office
b Showers, toilets & laundry
c Pub and restaurant
d RNLI, Inshore Lifeboat Centre
e Yacht stores

COWES HARBOUR
SERVICES
MOORINGS · BOATYARD · STORAGE · FUEL · COMMERCIAL
SEASONAL BOATYARD OFFERS
Sail and Power, Vessels to 40 Tonnes
Extensive Yard Facilities
Boat Maintenance Services
All-Tide Lifting, Choice of Storage Options
Secure Year-Round Under Cover Storage
Chandlery and Onsite Trades
Just 3 Minutes from Red Funnel East Cowes
Cowes Harbour Services, Kingston Wharf
East Cowes, Isle of Wight PO32 6JS
01983 299385 | boatyard.chc@cowes.co.uk
cowesharbourservices.co.uk

TOWN QUAY

Associated British Ports
Town Quay, Southampton, SO14 2AQ
Tel: 02380 234397 Mobile: 07764 293588
Email: info@townquay.com www.townquay.com

VHF Ch 80
ACCESS H24

In the heart of Southampton, Town Quay is walking distance from the City's cultural quarter, West Quay Shopping Centre and a variety of restaurants, bars and theatres making the marina a vibrant place to stay all year round.

The marina is accessible at all states of the tide and the reception is open 24 hours a day with free drinks and wi-fi. Free cycle hire and use of a gas BBQ on the 'chill out' deck is available.

Located on the eastern shores of Southampton Water, Town Quay offers unrivalled views of Southampton's busy maritime activity and direct access to the world famous cruising and racing waters of the Solent.

FACILITIES AT A GLANCE
WC

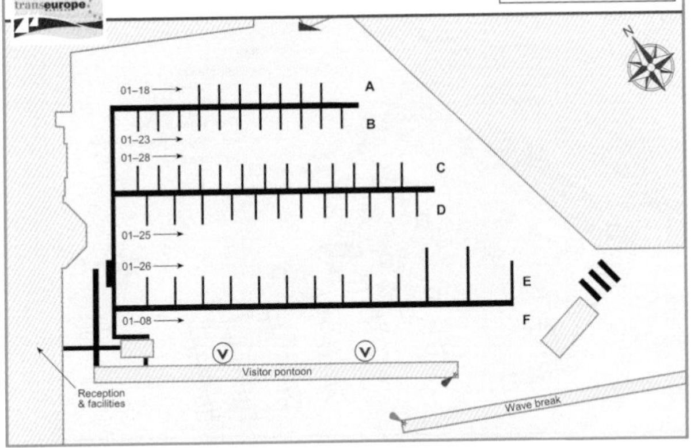
transeurope
01–18
01–23
01–28
01–25
01–26
01–08
A
B
C
D
E
F
Visitor pontoon
Wave break
Reception & facilities

OCEAN VILLAGE MARINA

Ocean Village Marina
2 Channel Way, Southampton, SO14 3TG
Tel: 023 8022 9385
Email: oceanvillage@mdlmarinas.co.uk
www.oceanvillagemarina.co.uk

VHF Ch 80
ACCESS H24

The entrance to Ocean Village Marina lies on the port side of the River Itchen, just before the Itchen Bridge. With the capacity to accommodate large yachts and tall ships, the marina, accessible 24 hours a day, is a renowned home for international yacht races.

Situated at the heart of an exciting new waterside development incorporating shops, a cinema, restaurants, a multi-storey car park and a £50m luxury spa hotel complex, Ocean Village offers a vibrant atmosphere for all visitors.

FACILITIES AT A GLANCE
WC

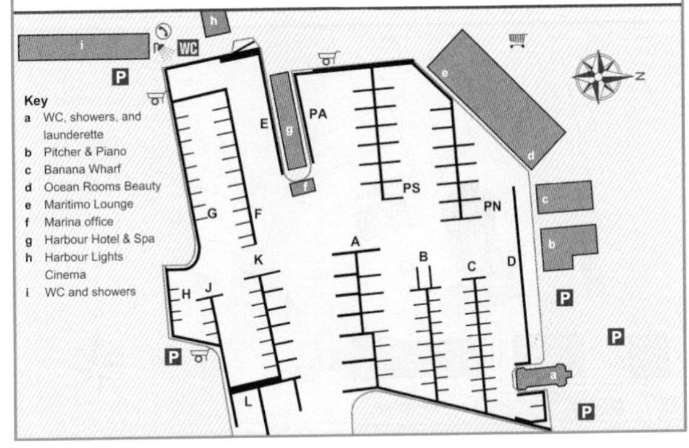
Key
a WC, showers, and launderette
b Pitcher & Piano
c Banana Wharf
d Ocean Rooms Beauty
e Maritimo Lounge
f Marina office
g Harbour Hotel & Spa
h Harbour Lights Cinema
i WC and showers
PA
PS
PN
A
B
C
D
E
F
G
H
J
K
L

SHAMROCK QUAY

Shamrock Quay
William Street, Northam, Southampton, Hants, SO14 5QL
Tel: 023 8022 9461
Email: shamrockquay@mdlmarinas.co.uk
www.shamrockquay.co.uk

VHF Ch 80
ACCESS H24

Shamrock Quay, lying upstream of the Itchen Bridge on the port hand side, offers excellent facilities to yachtsmen. It also benefits from being accessible and manned 24/7 a day. On-site there is a 75-ton travel hoist and a 47-ton boat mover, and for dining out a fully licenced restaurant/bar and a café.

The city centre is about two miles away, where among the numerous attractions are the Medieval Merchant's House in French Street, the Southampton City Art Gallery and the SeaCity Museum in the Civic Centre.

FACILITIES AT A GLANCE
WC

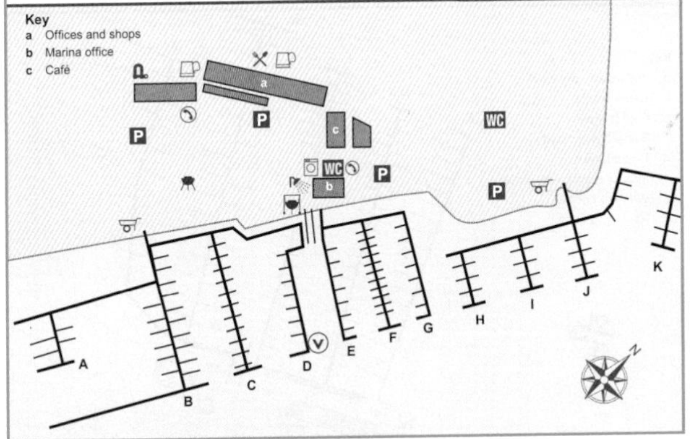
Key
a Offices and shops
b Marina office
c Café
A
B
C
D
E
F
G
H
I
J
K

KEMPS QUAY

Kemp's Shipyard Ltd
Quayside Road, Southampton, SO18 1BZ
Tel: 023 8063 2323 Fax: 023 8022 6002
Email: enquiries@kempsquay.com

VHF
ACCESS HW±3.5

At the head of the River Itchen on the starboard side is Kemps Quay, a family-run marina with a friendly, old-fashioned feel. Accessible only 3½ hrs either side of HW, it has a limited number of deep water berths, the rest being half tide, drying out to soft mud. Its restricted access is, however, reflected in the lower prices.

Although situated on the outskirts of Southampton, a short bus or taxi ride will soon get you to the city centre. Besides a nearby BP Garage selling bread and milk, the closest supermarkets can be found in Bitterne Shopping Centre, which is five minutes away by bus.

FACILITIES AT A GLANCE
WC

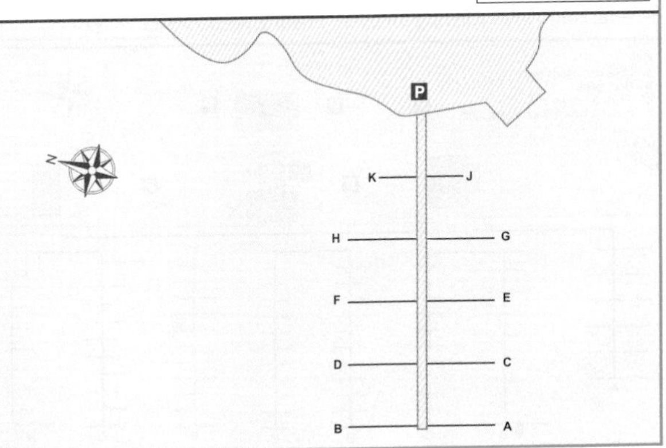
K
J
H
G
F
E
D
C
B
A

SAXON WHARF

Saxon Wharf
Lower York Street, Northam, Southampton, SO14 5QF
Tel: 023 8033 9490
Email: saxonwharf@mdlmarinas.co.uk
www.saxonwharfmarina.co.uk

VHF Ch 80
ACCESS H24

Placed on the River Itchen in Saxon Wharf is a marine service specifically designed for the superyacht market. With a 200-ton boat hoist and heavy duty pontoons, Saxon Wharf is the ideal location for large vessels in need of secure, quick turnaround lift-outs, repair work or even full-scale refits.

With a Dry Stack facility boasting the largest capacity forklift truck in the UK, Saxon Wharf can now dry stack boats of up to 13m LOA. There is also ample storage ashore and 24-hour security. Shamrock Quay, where there are bars and restaurants, is within 300 metres of this location.

FACILITIES AT A GLANCE

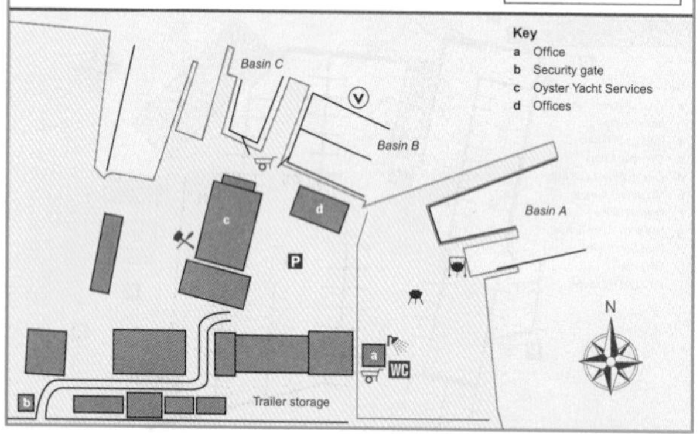
Key
a Office
b Security gate
c Oyster Yacht Services
d Offices
Basin C
Basin B
Basin A
Trailer storage
N

HAMBLE POINT MARINA

Hamble Point Marina
School Lane, Hamble, Southampton, SO31 4NB
Tel: 023 8045 2464
Email: hamblepoint@mdlmarinas.co.uk
www.hamblepointmarina.co.uk

VHF Ch 80
ACCESS H24

Situated virtually opposite Warsash, this is the first marina you will come to on the western bank of the Hamble. Accommodating yachts and power boats up to 30m in length, it offers easy access to the Solent.

The marina boasts extensive facilities including 137 dry stack berths for motorboats up to 10m and over 50 tenants who provide boat-owners with a wide range of marine services from boat repairs to electrical work. Hamble Point is within a 20-minute walk of Hamble Village, where there are a plethora of pubs and restaurants on offer.

FACILITIES AT A GLANCE

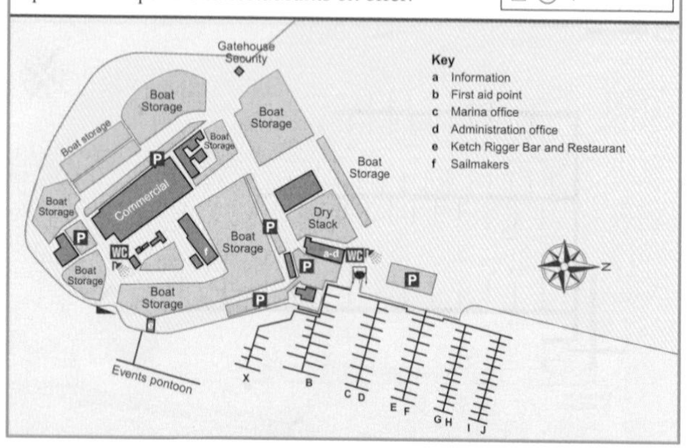
Key
a Information
b First aid point
c Marina office
d Administration office
e Ketch Rigger Bar and Restaurant
f Sailmakers
Gatehouse Security
Boat Storage
Commercial
Dry Stack
Events pontoon

PORT HAMBLE MARINA

Port Hamble Marina
Satchell Lane, Hamble, Southampton, SO31 4QD
Tel: 023 8045 2741
Email: porthamble@mdlmarinas.co.uk
www.porthamblemarina.co.uk

VHF Ch 80
ACCESS H24

Port Hamble Marina is situated on the River Hamble right in the heart of the South Coast's sailing scene. With thousands of visitors every year, this busy marina is popular with racing enthusiasts and cruising vessels looking for a vibrant atmosphere. The picturesque Hamble village, with its inviting pubs and restaurants, is only a few minutes walk away.

On site, Port Hamble also offers excellent amenities including luxurious male and female facilities and boutique style shower rooms for members. Banana Wharf bar and restaurant provides the perfect spot to socialise by the water, whilst petrol and diesel is available 7/7. Locally there are several companies catering for every boating need.

FACILITIES AT A GLANCE

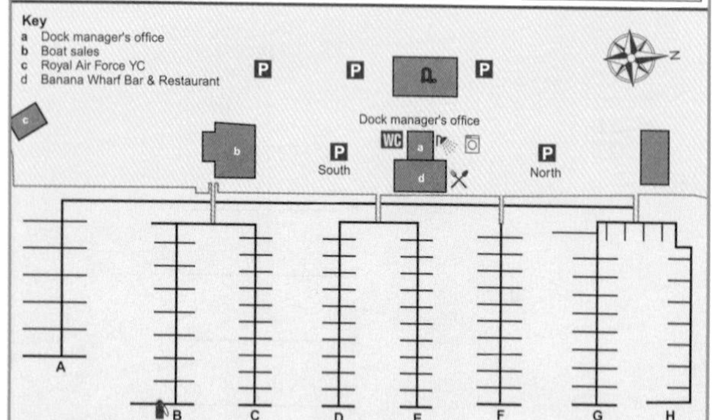
Key
a Dock manager's office
b Boat sales
c Royal Air Force YC
d Banana Wharf Bar & Restaurant
Dock manager's office
South
North

MERCURY YACHT HARBOUR

Mercury Yacht Harbour and Holiday Park
Satchell Lane, Hamble, Southampton, SO31 4HQ
Tel: 023 8045 5994
Email: mercury@mdlmarinas.co.uk
www.mercuryyachtharbour.co.uk

VHF Ch 80
ACCESS H24

Mercury Yacht Harbour is set in a picturesque and sheltered wooded site where the shallow waters of Badnam Creek join the River Hamble. Enjoying deep water at all states of the tide, it accommodates yachts up to 24m LOA and boasts an extensive array of facilities.

For a good meal look no further than the Gaff Rigger bar and restaurant, whose roof terrace offers striking views over the water. Hamble Village with plenty more pubs, restaurants and shops is only a 20-minute walk away.

FACILITIES AT A GLANCE

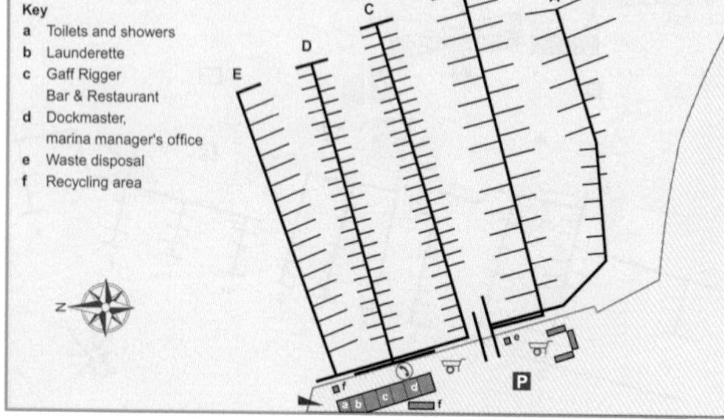
Key
a Toilets and showers
b Launderette
c Gaff Rigger Bar & Restaurant
d Dockmaster, marina manager's office
e Waste disposal
f Recycling area

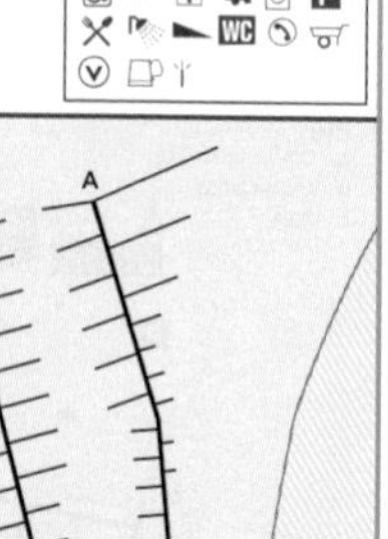

UNIVERSAL MARINA

Universal Marina
Crableck Lane, Sarisbury Green, Southampton, SO31 7ZN
Tel: 01489 574272 Fax: 01489 574273
Email: info@universalmarina.co.uk

VHF	Ch 80
ACCESS	H24

Universal Marina is one of the few remaining independent marinas offering south coast moorings. Universal Marina's unique location is unbeatable, tucked in between the oak trees on the East Bank of the Hamble where 68 acres of natural wildlife and marshlands surrounds the busy and friendly marina. Positioned only minutes off the M27, it is one of the most accessible marinas on the south coast. The 250 berth complex, features all the latest facilities required by the modern day boat owner, recently upgraded pontoons, power, water and wifi available to each berth. Visitors are welcome & although there are no dedicated visitor berths these are available by prior arrangement.

FACILITIES AT A GLANCE
WC

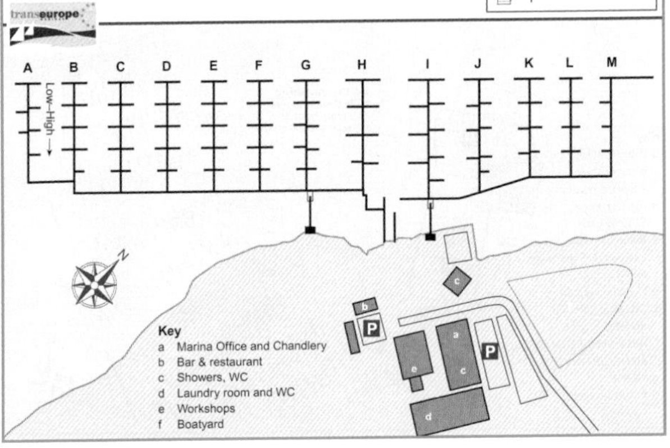
transeurope
A
B
C
D
E
F
G
H
I
J
K
L
M
Low–High
Key
a Marina Office and Chandlery
b Bar & restaurant
c Showers, WC
d Laundry room and WC
e Workshops
f Boatyard

SWANWICK MARINA

Swanwick Marina
Swanwick, Southampton, Hampshire, SO31 1ZL
Tel: 01489 884081
Email: swanwick@premiermarinas.com
www.premiermarinas.com

VHF	Ch 80
ACCESS	H24

Nestling on the picturesque eastern bank of the River Hamble, Swanwick Marina offers 24/7 access to the cruising grounds of the Solent. There is excellent access by road and a variety of walks, pubs and restaurants close by.

Welcoming to boats of all sizes, the marina has over 320 berths, and a choice of water-berthing and dry-stack. There is a café, secure parking and construction is underway for an new Pavilion building, which will house boat sales and marine businesses; Swanwick also boasts a fully-serviced boatyard, a self-store and a mix of marine services, including a chandlery

FACILITIES AT A GLANCE
WC

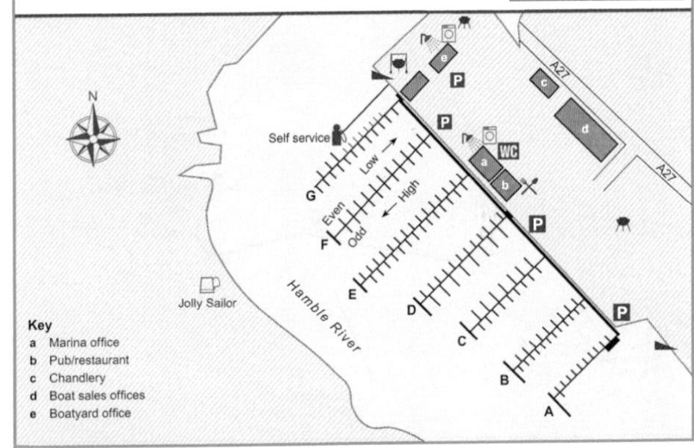
N
A27
A27
Self service
Low
High
Even
Odd
G
F
E
D
C
B
A
Jolly Sailor
Hamble River
WC
Key
a Marina office
b Pub/restaurant
c Chandlery
d Boat sales offices
e Boatyard office

DEACONS MARINA & BOATYARD

Deacons Marina & Boatyard
Bridge Road, Bursledon, Hampshire, SO31 8AZ
Tel: 02380 402253
Email: deacons@boatfolk.co.uk
www.boatfolk.co.uk/deaconsmarina

VHF	Ch 80
ACCESS	H24

Deacons Marina and Boatyard is situated on a sheltered bank on the western side of the Hamble River, a stones throw from Bursledon, with 130 berths afloat and space for 160 boats ashore.

The marina boasts friendly and knowledgeable staff, skilled in boat repair and maintenance and offers craneage facilities for lifting boats up to 20t and 45ft in length. All supported by marine on-site businesses to provide a fully serviced one stop facility for boat owners.

FACILITIES AT A GLANCE

WC

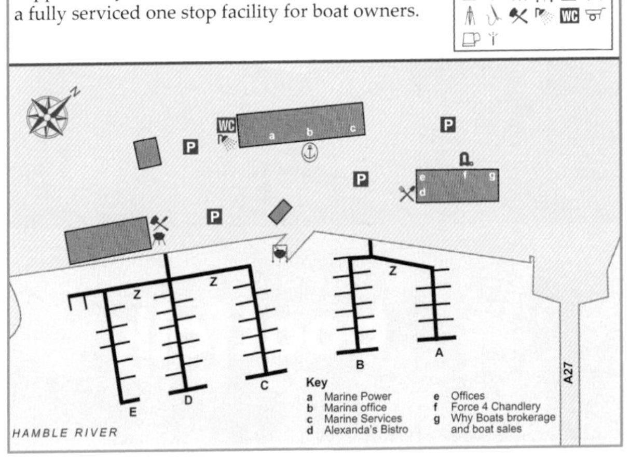
WC
a
b
c
d
e
f
g
Z
Z
Z
E
D
C
B
A
A27
HAMBLE RIVER
Key
a Marine Power
b Marina office
c Marine Services
d Alexanda's Bistro
e Offices
f Force 4 Chandlery
g Why Boats brokerage and boat sales

RYDE LEISURE HARBOUR

Ryde Harbour
The Esplanade, Ryde, Isle of Wight, PO33 1JA
Tel: 01983 613879 Fax: 01983 613903
www.rydeharbour.com Email: ryde.harbour@iow.gov.uk

VHF	Ch 80
ACCESS	HW±2

Known as the 'gateway to the Island', Ryde, with its elegant houses and abundant shops, is among the Isle of Wight's most popular resorts. Its well-protected harbour is conveniently close to the exceptional beaches as well as to the town's restaurants and amusements.

Drying to 2.5m and therefore only accessible to yachts that can take the ground, the harbour accommodates 90 resident boats as well as up to 75 visiting yachts. Fin keel yachts may dry out on the harbour wall.

Ideal for family cruising, Ryde offers a wealth of activities, ranging from ten pin bowling and ice skating to crazy golf and tennis.

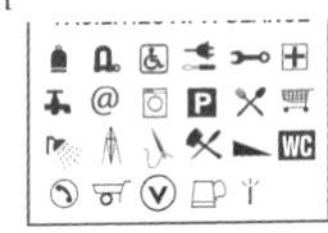
WC

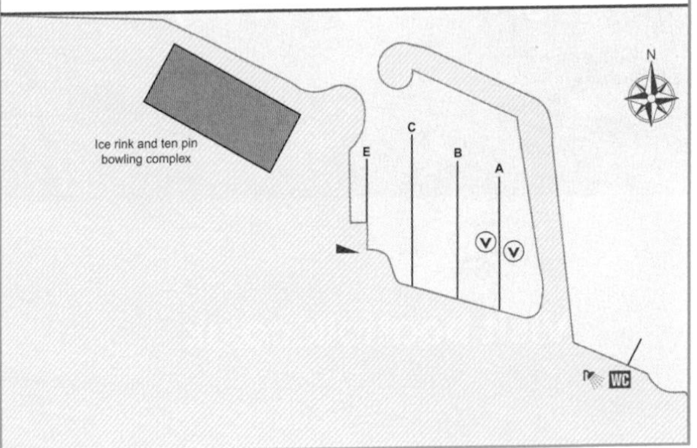
Ice rink and ten pin bowling complex
N
E
C
B
A
WC

BEMBRIDGE HARBOUR

Bembridge Harbour
Harbour Office, The Duver, St Helens, Ryde
Isle of Wight, PO33 1YB
Tel: 01983 872828 Fax: 01983 872922
Email: thorpemalcolm@btconnect.com
www.bembridgeharbour.co.uk

VHF	Ch 80
ACCESS	HW±2.5

Bembridge is a compact, pretty harbour whose entrance, although restricted by the tides (recommended entry for a 1.5m draught is 2½hrs before HW), is well sheltered in all but north north easterly gales. Offering excellent sailing clubs, beautiful beaches and fine restaurants, this Isle of Wight port is a first class haven with plenty of charm. With approximately 120 new visitors' berths on the Duver Marina pontoons, which can now be booked online, the marina at St Helen's Quay at the western end of the harbour is now allocated to annual berth holders only.

FACILITIES AT A GLANCE

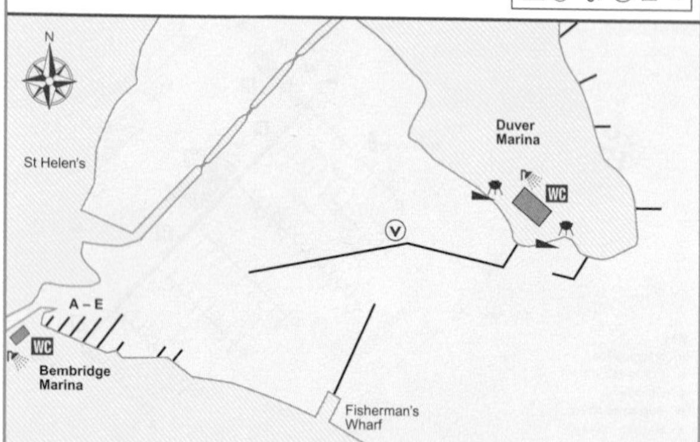
N
St Helen's
Duver Marina
WC
A – E
WC
Bembridge Marina
Fisherman's Wharf

HASLAR MARINA

Haslar Marina
Haslar Road, Gosport, Hampshire, PO12 1NU
Tel: 023 9260 1201
Email: haslar@boatfolk.co.uk
www.boatfolk.co.uk/haslarmarina

VHF	Ch 80
ACCESS	H24

This purpose-built marina lies to port on the western side of Portsmouth Harbour entrance and is easily recognised by its prominent lightship incorporating a bar and restaurant. Accessible at all states of the tide, Haslar's extensive facilities do not however include fuel, the nearest is at the Gosport Marina only a few cables north. Within close proximity is the Royal Navy Submarine Museum and the Museum of Naval Firepower 'Explosion' both worth a visit.

FACILITIES AT A GLANCE

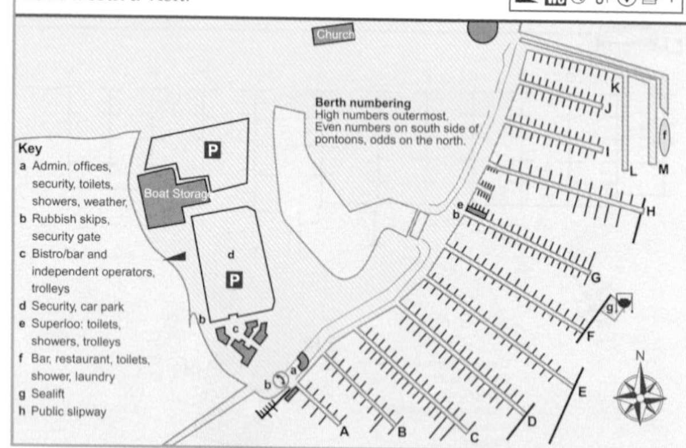
Church
Berth numbering
High numbers outermost.
Even numbers on south side of pontoons, odds on the north.
Key
a Admin. offices, security, toilets, showers, weather
b Rubbish skips, security gate
c Bistro/bar and independent operators, trolleys
d Security, car park
e Superloo: toilets, showers, trolleys
f Bar, restaurant, toilets, shower, laundry
g Sealift
h Public slipway
P
Boat Storage
A
B
C
D
E
F
G
H
I
J
K
L
M

GOSPORT MARINA

Gosport Marina
Mumby Road, Gosport, Hampshire, PO12 1AH
Tel: 023 9252 4811
Email: gosport@premiermarinas.com
www.premiermarinas.com

VHF Ch 80
ACCESS H24

Gosport Marina is located at the mouth of Portsmouth Harbour, just minutes from the eastern edge of the Solent. Family-friendly and popular with cruisers and racers alike, Gosport offers 24/7 access to open water and a choice of both traditional wet-berth and dry-stack, plus luxury facilities and a self-store. The marina's specialist boatyard, Endeavour Quay, is fully-equipped to lift and service boats up to 40m, with three large boatsheds and a wide variety of marine service tenants, including a dedicated chandlery. When it is time to relax, the Boat House Café is the perfect spot to unwind.

FACILITIES AT A GLANCE

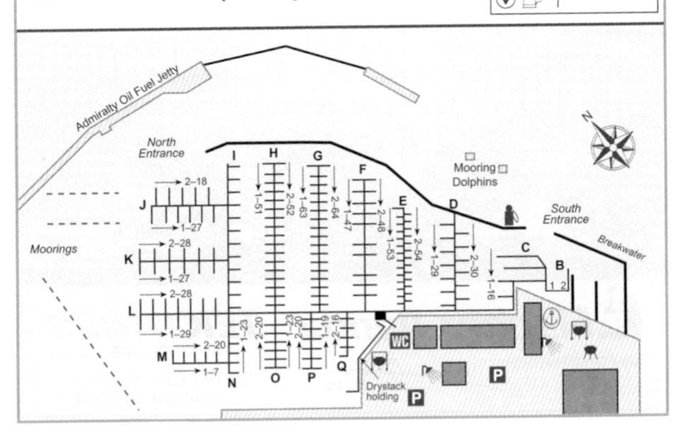
Admiralty Oil Fuel Jetty
North Entrance
Mooring Dolphins
South Entrance
Breakwater
Moorings
Drystack holding
I
H
G
F
E
D
C
B
J
K
L
M
N
O
P
Q
WC
P

ROYAL CLARENCE MARINA

Royal Clarence Marina, Royal Clarence Yard
Weevil Lane, Gosport, Hampshire PO12 1AX
Tel: 02392 523523
Email: info@royalclarencemarina.org
www.royalclarencemarina.org

VHF Ch 80
ACCESS H24

Royal Clarence Marina enjoys a unique setting, with the former Royal Navy victualling yard as its backdrop. Only five minutes from the entrance of Portsmouth Harbour, this Transeurope marina lies within a deep-water basin giving 24/7 access for a draft of up to 4.5m. This remarkably peaceful and calm marina has wide pontoon spacing, little tidal flow, and exceptional protection from the swell produced in Portsmouth Harbour. The new facilities, now just opposite the marina, have been described as the best on the South Coast and with two popular restaurants onsite, makes a great location for visitors.

FACILITIES AT A GLANCE

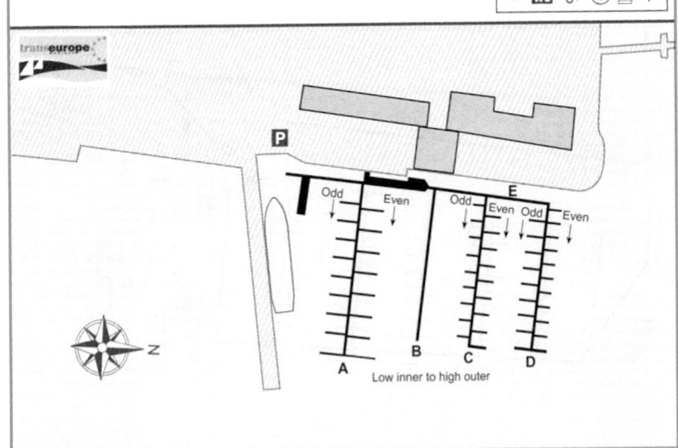
transeurope
P
Odd
Even
Odd
Even
Odd
Even
E
A
B
C
D
Low inner to high outer
N

PORT SOLENT MARINA

Port Solent Marina
South Lockside, Portsmouth, PO6 4TJ
Tel: 023 9221 0765
Email: portsolent@premiermarinas.com
www.premiermarinas.com

VHF Ch 80
ACCESS H24

Port Solent Marina is located to the north east of Portsmouth Harbour, not far from the Roman Portchester Castle with its Norman keep and church.

Accessible via a 24/7 lock, this purpose built marina offers a secure and sheltered berthing for yachts and motorboats alike. There is a complete range of facilities including a fully serviced boatyard with a sizable boatshed and a self-store. The Boardwalk comprises of an array of shops and restaurants, while close by is a gym and a large cinema complex.

FACILITIES AT A GLANCE

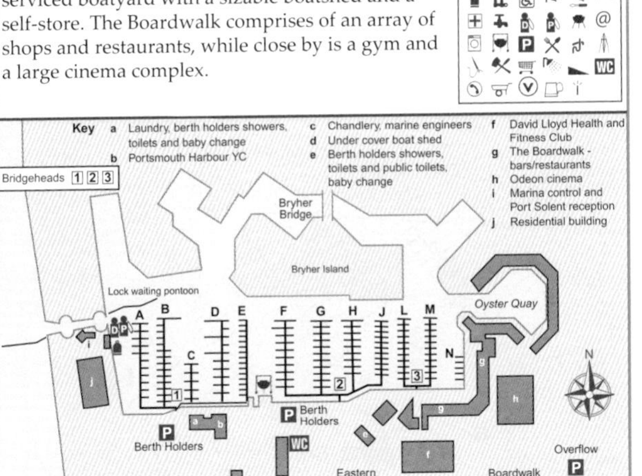
Key
a Laundry, berth holders showers, toilets and baby change
b Portsmouth Harbour YC
c Chandlery, marine engineers
d Under cover boat shed
e Berth holders showers, toilets and public toilets, baby change
f David Lloyd Health and Fitness Club
g The Boardwalk - bars/restaurants
h Odeon cinema
i Marina control and Port Solent reception
j Residential building
Bridgeheads 1 2 3
Bryher Bridge
Bryher Island
Lock waiting pontoon
Oyster Quay
Berth Holders
Western Boat Park
Slipway Compound
Eastern Boat Park
Boardwalk
Overflow
N

WICORMARINE YACHT HAVEN

WicorMarine Yacht Haven
Cranleigh Road, Portchester, Hampshire, PO16 9DR
Tel: 01329 237112 Fax: 01329 248595
Email: inbox@wicormarine.co.uk

VHF
ACCESS H24

WicorMarine Yacht Haven is situated in the picturesque upper reaches of Portsmouth Harbour away from the hustle and bustle. The walk-ashore pontoons and traditional mid-river berths offer an affordable alternative to busy marinas and are only 30 mins from the harbour entrance.

An excellent range of boatyard facilities including a 12T boat hoist, undercover storage, H24 showers and toilets, diesel, fresh water, on site repair services and a chandlery complete with Calor Gas and Camping Gaz exchange.

The popular, licensed Salt Cafe is open to visitors all year round where you can take in the stunning views of the harbour from the waterfront deck.

FACILITIES AT A GLANCE

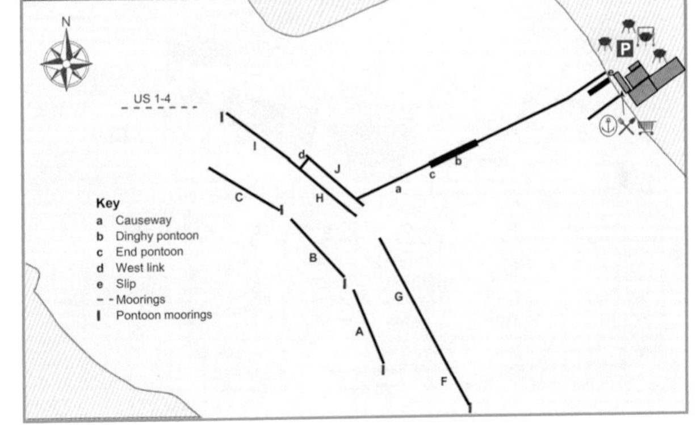
N
US 1-4
Key
a Causeway
b Dinghy pontoon
c End pontoon
d West link
e Slip
- - Moorings
I Pontoon moorings
A
B
C
F
G
H
I
J

SOUTHSEA MARINA

Southsea Marina
Fort Cumberland Road, PO4 9RJ
Tel: 02392 822719
Email: southsea@premiermarinas.com
www.premiermarinas.com

VHF	Ch 80
ACCESS	HW±3

Tucked just inside Eastney Peninsula, in the quieter reaches of Langstone Harbour, Southsea Marina is perfectly located for exploring the Solent. A family-friendly, working marina, ideal for sailing and motor cruisers alike; Southsea offers value, first-class facilities and a personal service that includes a 24/7 manned reception. Access to and from the marina is via a tidal gate up to 3 hours either side of HW. There is full-service boatyard, ample storage ashore and a mix of onsite marine services. Two onsite restaurants offer a choice of Indian and home-style cuisine.

FACILITIES AT A GLANCE

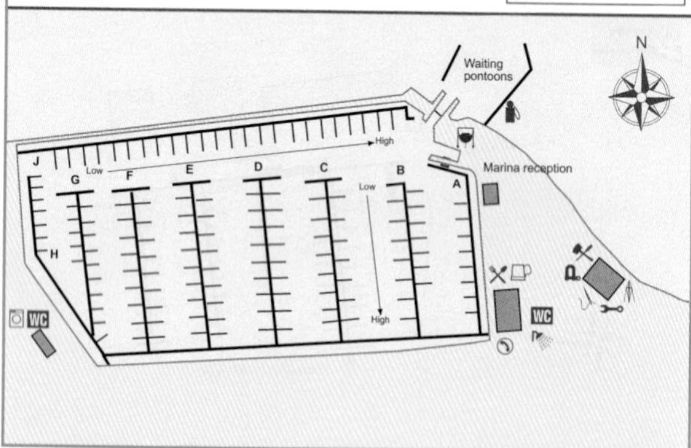
Waiting pontoons
N
High
Low
J
G
F
E
D
C
B
A
H
Marina reception
WC

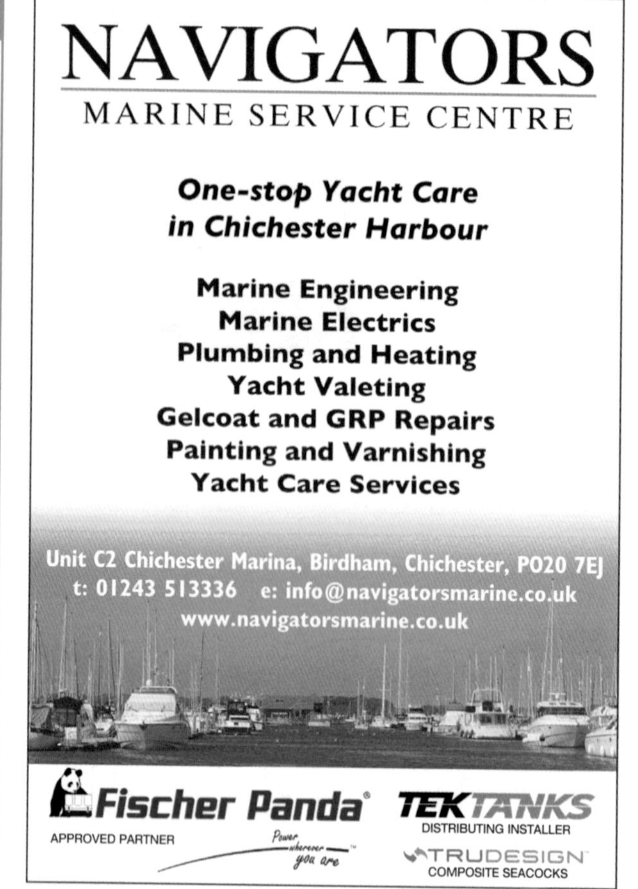
NAVIGATORS
MARINE SERVICE CENTRE
One-stop Yacht Care in Chichester Harbour
Marine Engineering
Marine Electrics
Plumbing and Heating
Yacht Valeting
Gelcoat and GRP Repairs
Painting and Varnishing
Yacht Care Services
Unit C2 Chichester Marina, Birdham, Chichester, PO20 7EJ
t: 01243 513336 e: info@navigatorsmarine.co.uk
www.navigatorsmarine.co.uk
Fischer Panda
APPROVED PARTNER
Power wherever you are
TEKTANKS
DISTRIBUTING INSTALLER
TRUDESIGN
COMPOSITE SEACOCKS

SPARKES MARINA

Sparkes Marina
38 Wittering Road, Hayling Island, Hampshire, PO11 9SR
Tel: 023 9246 3572
Email: sparkes@mdlmarinas.co.uk www.sparkesmarina.co.uk

VHF	Ch 80
ACCESS	H24

Just inside the entrance to Chichester Harbour, on the eastern shores of Hayling Island, lies Sparkes Marina. Its facilities include 24-hour showers and toilets, a laundry room, an office/ reception, and Drift: Hayling Island bar and restaurant.

In addition to its berthing and marina services, Sparkes has many skilled professionals on site, including specialists in engineering, outboard engines, glass fibre repairs, rigging, sails and covers, marine carpentry, electrical, boat management and valeting. There is storage ashore for over 200 boats and a 40 ton mobile crane (lifting capacity 15 tons).

FACILITIES AT A GLANCE

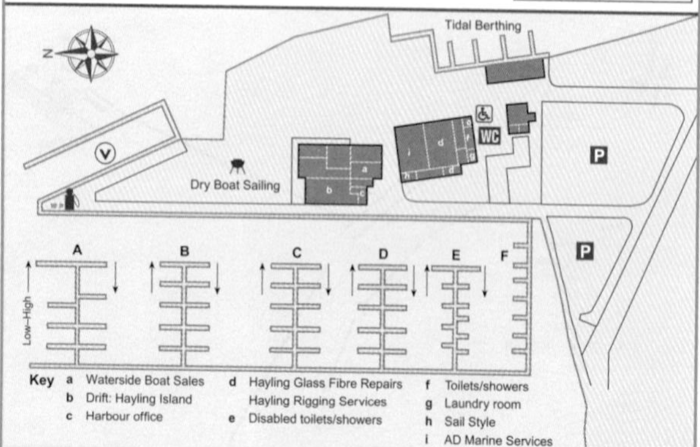
Tidal Berthing
N
Dry Boat Sailing
WC
P
A
B
C
D
E
F
Low–High
Key
a Waterside Boat Sales
b Drift: Hayling Island
c Harbour office
d Hayling Glass Fibre Repairs
Hayling Rigging Services
e Disabled toilets/showers
f Toilets/showers
g Laundry room
h Sail Style
i AD Marine Services

NORTHNEY MARINA

Northney Marina
Northney Road, Hayling Island, Hampshire, PO11 0NH
Tel: 023 9246 6321
Email: northney@mdlmarinas.co.uk
www.northneymarina.co.uk

VHF	Ch 80
ACCESS	H24

Situated in Chichester Harbour, Northney Marina is set on the northern shore of Hayling Island in the well-marked Sweare Deep Channel, which branches off to port almost at the end of Emsworth Channel. This 228-berth marina offers excellent boatyard facilities, a provisions store and laundry area, diesel, a slipway, Salt Shack Café, wi-fi and 24/7 staff cover. There is also an events area for rallies.

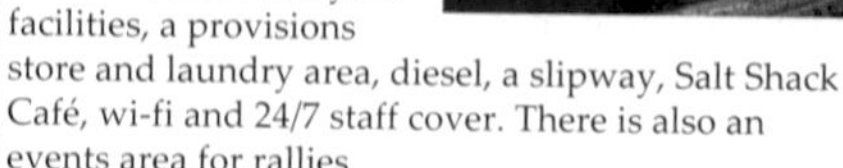

FACILITIES AT A GLANCE

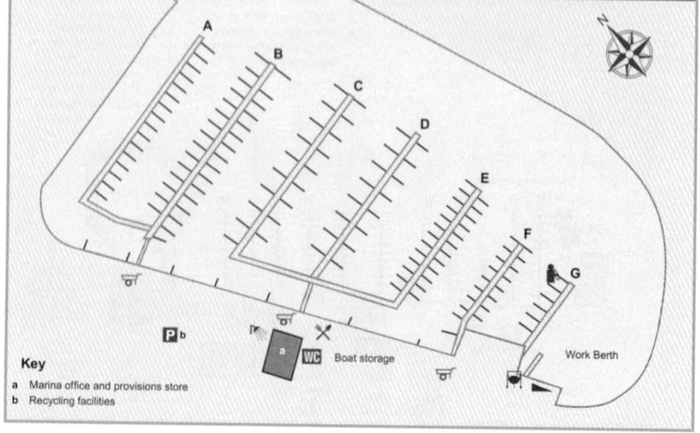
N
A
B
C
D
E
F
G
WC
Boat storage
Work Berth
Key
a Marina office and provisions store
b Recycling facilities

EMSWORTH YACHT HARBOUR

Emsworth Yacht Harbour Ltd
Thorney Road, Emsworth, Hants, PO10 8BP
Tel: 01243 377727 Fax: 01243 373432
Email: info@emsworth-marina.co.uk
www.emsworth-marina.co.uk

VHF
ACCESS HW±2

Accessible about one and a half to two hours either side of high water, Emsworth Yacht Harbour is a sheltered site, offering good facilities to yachtsmen.

Created in 1964 from a log pond, the marina is within easy walking distance of the pretty little town of Emsworth, which boasts at least 10 pubs, several high quality restaurants and two well-stocked convenience stores.

FACILITIES AT A GLANCE

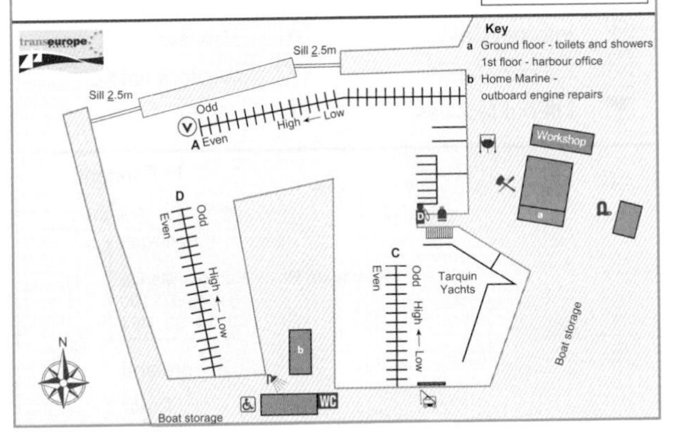
transeurope
Sill 2.5m
Sill 2.5m
Key
a Ground floor - toilets and showers
1st floor - harbour office
b Home Marine -
outboard engine repairs
Odd
Even
A
High
Low
Workshop
D
C
Tarquin Yachts
Boat storage
Boat storage
N
WC

CHICHESTER MARINA

Chichester Marina
Birdham, Chichester, West Sussex, PO20 7EJ
Tel: 01243 512731
Email: chichester@premiermarinas.com
www.premiermarinas.com

VHF Ch 80
ACCESS HW±5

Set in the beauty of a natural harbour and picturesque countryside makes Chichester Marina a wonderful destination.

This locked, family-friendly, marina is home to over 1000 berths and combines luxury facilities with serene surroundings. It is also home to a friendly yacht club and The Boat House Café, which offers bistro dining. The marina also boasts a fully-serviced boatyard including two boat hoists, a slipway and boat storage, plus a wide variety of marine services.

Chichester is only about a five minute bus or taxi ride away and has a host of places of interest, the most notable being the cathedral.

FACILITIES AT A GLANCE

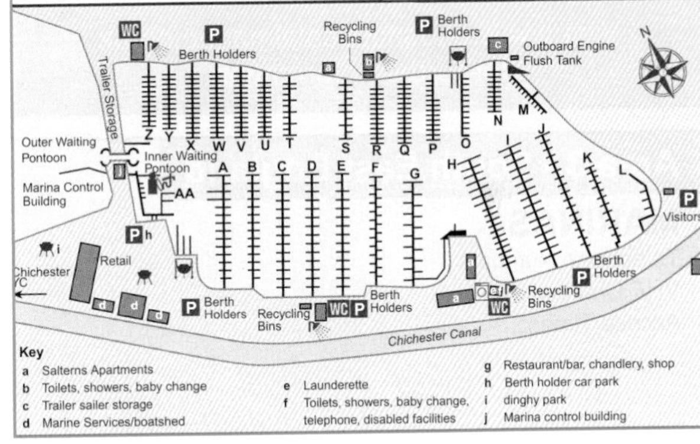
Trailer Storage
Berth Holders
Recycling Bins
Berth Holders
Outboard Engine Flush Tank
Outer Waiting Pontoon
Inner Waiting Pontoon
Marina Control Building
Retail
Chichester YC
Berth Holders
Recycling Bins
Berth Holders
Recycling Bins
Berth Holders
Visitors
Chichester Canal
Key
a Salterns Apartments
b Toilets, showers, baby change
c Trailer sailer storage
d Marine Services/boatshed
e Launderette
f Toilets, showers, baby change, telephone, disabled facilities
g Restaurant/bar, chandlery, shop
h Berth holder car park
i dinghy park
j Marina control building

BIRDHAM POOL MARINA

Birdham Pool Marina
Birdham Pool, Chichester, Sussex
Tel: 01243 512310 Fax: 01243 513163
Email: mikebraidley@castlemarinas.co.uk

VHF Ch 80
ACCESS HW±3·5

Birdham Pool is the UK's oldest marina, with a charm not found elsewhere. A recent programme of improvements sees 28 new berths, a new 30-ton crane to lift out boats up to 50ft long and 2m draft, brand new facilities block and new Marine Trades Centre showcasing traditional skilled craftsmen. Visitors will not be disappointed by the unique and picturesque setting, with views across the South Downs as well as Chichester Harbour. Access is HW±3·5, via a dredged channel marked by starboard hand piles.

FACILITIES AT A GLANCE

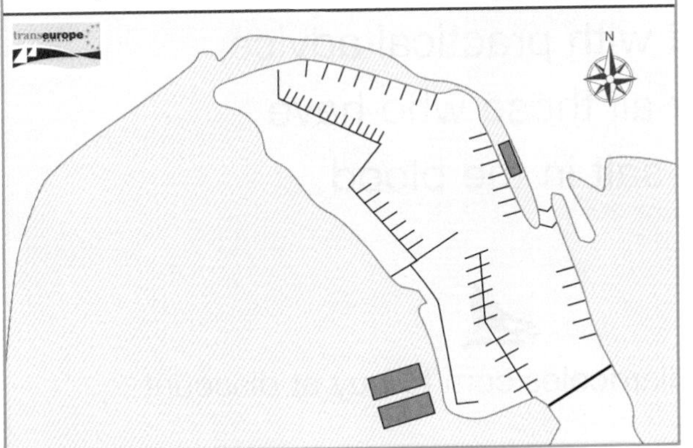
transeurope
N

SKIPPER
TIPS
FOR EVERY DAY
365
WAYS TO GET THE MOST FROM YOUR SAILING
FRIDTJOF GUNKEL

SOUTH EAST ENGLAND – Selsey Bill to North Foreland

Key to Marina Plans symbols

Bottled gas	Parking
Chandler	Pub/Restaurant
Disabled facilities	Pump out
Electrical supply	Rigging service
Electrical repairs	Sail repairs
Engine repairs	Shipwright
First Aid	Shop/Supermarket
Fresh Water	Showers
Fuel - Diesel	Slipway
Fuel - Petrol	Toilets
Hardstanding/boatyard	Telephone
Internet Café	Trolleys
Laundry facilities	Visitors berths
Lift-out facilities	Wi-Fi

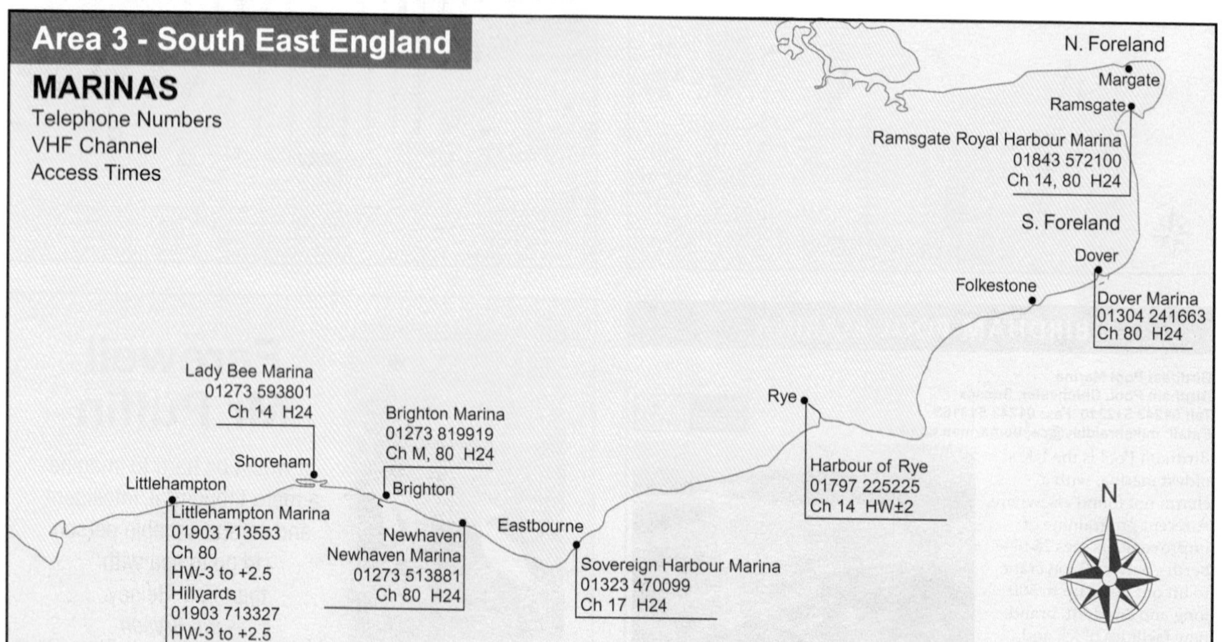
Area 3 - South East England
MARINAS
Telephone Numbers
VHF Channel
Access Times
N. Foreland
Margate
Ramsgate
Ramsgate Royal Harbour Marina
01843 572100
Ch 14, 80 H24
S. Foreland
Dover
Folkestone
Dover Marina
01304 241663
Ch 80 H24
Rye
Harbour of Rye
01797 225225
Ch 14 HW±2
Lady Bee Marina
01273 593801
Ch 14 H24
Shoreham
Brighton Marina
01273 819919
Ch M, 80 H24
Brighton
Littlehampton
Littlehampton Marina
01903 713553
Ch 80
HW-3 to +2.5
Hillyards
01903 713327
HW-3 to +2.5
Newhaven
Newhaven Marina
01273 513881
Ch 80 H24
Eastbourne
Sovereign Harbour Marina
01323 470099
Ch 17 H24
N

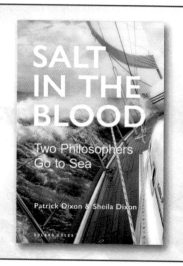

LITTLEHAMPTON MARINA

Littlehampton Marina
Ferry Road, Littlehampton, W Sussex
Tel: 01903 713553 Fax: 01903 732264
Email: sales@littlehamptonmarina.co.uk

VHF	Ch 80
ACCESS	HW-3 to +2.5

A typical English seaside town with funfair, promenade and fine sandy beaches, Littlehampton lies roughly midway between Brighton and Chichester at the mouth of the River Arun. It affords a convenient stopover for yachts either east or west bound, providing you have the right tidal conditions to cross the entrance bar with its charted depth of 0.7m. The marina lies about three cables above Town Quay and Fisherman's Quay, both of which are on the starboard side of the River Arun, and is accessed via a retractable footbridge that opens on request to the HM (note that you should contact him by 1630 the day before you require entry).

FACILITIES AT A GLANCE

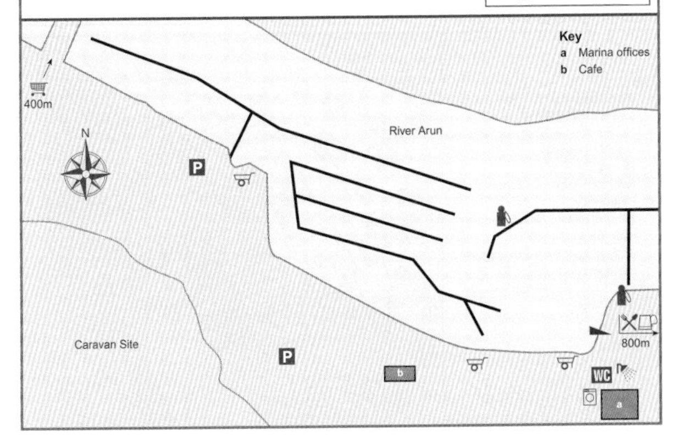
Key
a Marina offices
b Cafe
400m
River Arun
Caravan Site
800m
P
WC

LADY BEE MARINA

Lady Bee Marina
138-140 Albion Street, Southwick
West Sussex, BN42 4EG
Tel: 01273 593801 Fax: 01273 870349

VHF	Ch 14
ACCESS	H24

Shoreham, only five miles west of Brighton, is one of the South Coast's major commercial ports handling, among other products, steel, grain, tarmac and timber. On first impressions it may seem that Shoreham has little to offer the visiting yachtsman, but once through the lock and into the eastern arm of the River Adur, the quiet Lady Bee Marina, with its Spanish waterside restaurant, can make this harbour an interesting alternative to the lively atmosphere of Brighton Marina. Run by the Harbour Office, the marina meets all the usual requirements, although fuel is available in cans from Southwick garage or from Corral's diesel pump situated in the western arm.

FACILITIES AT A GLANCE

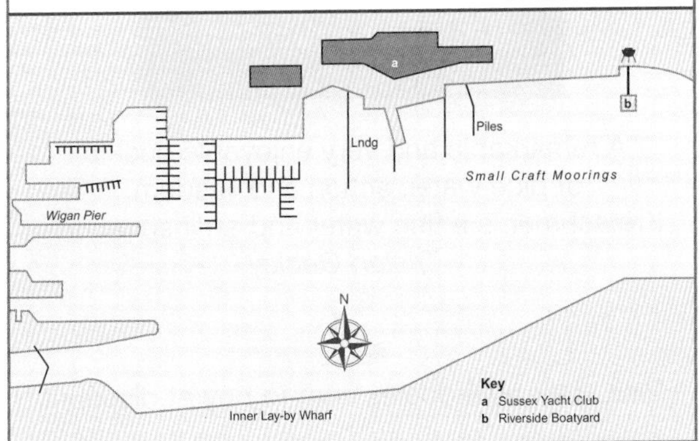
Lndg
Piles
Small Craft Moorings
Wigan Pier
Inner Lay-by Wharf
N
Key
a Sussex Yacht Club
b Riverside Boatyard

BRIGHTON MARINA

Brighton Marina
West Jetty, Brighton, East Sussex, BN2 5UP
Tel: 01273 819919
Email: brighton@premiermarinas.com
www.premiermarinas.com

VHF	Ch M, 80
ACCESS	H24

Perfect for yachts and motorboats alike, Brighton is the UK's largest marina with over 1300 berths and offers easy access to open water and a great starting point for exploring the South Coast or a trip to France.

Brighton Marina also boasts a modern and friendly yacht club that warmly welcomes members and visiting yachtsman. With a fully serviced, quality boatyard, storage ashore, luxury facilities and a comprehensive mix of tenant marine services including a chandlery, together these make Brighton Marina a great place to berth and relax, and carry out major or routine boat repairs on the South Coast.

FACILITIES AT A GLANCE

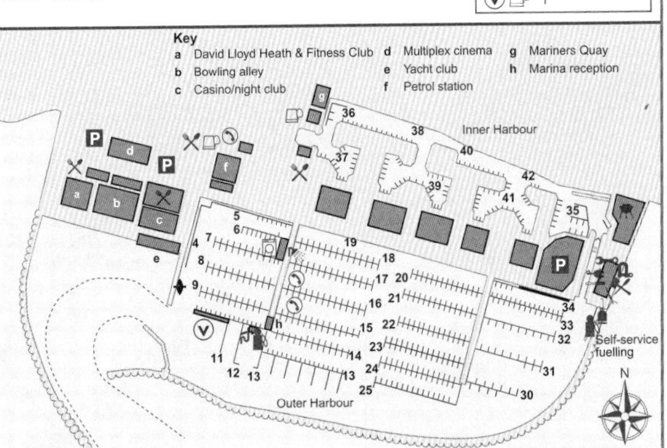
Key
a David Lloyd Heath & Fitness Club
b Bowling alley
c Casino/night club
d Multiplex cinema
e Yacht club
f Petrol station
g Mariners Quay
h Marina reception
Inner Harbour
Outer Harbour
Self-service fuelling
N

NEWHAVEN MARINA

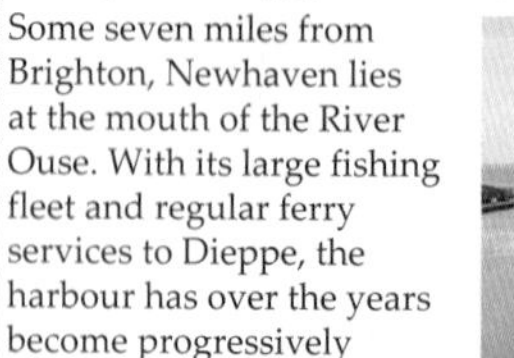

Newhaven Marina
The Yacht Harbour, Fort Road, Newhaven
East Sussex, BN9 9BY
Tel: 01273 513881
Email: john.stirling@newhavenmarina.co.uk

VHF	Ch 80
ACCESS	H24

Some seven miles from Brighton, Newhaven lies at the mouth of the River Ouse. With its large fishing fleet and regular ferry services to Dieppe, the harbour has over the years become progressively commercial, therefore care is needed to keep clear of large vessels under manoeuvre.
The marina lies approximately quarter of a mile from the harbour entrance on the west bank and was recently dredged to allow full tidal access except on LWS.

FACILITIES AT A GLANCE

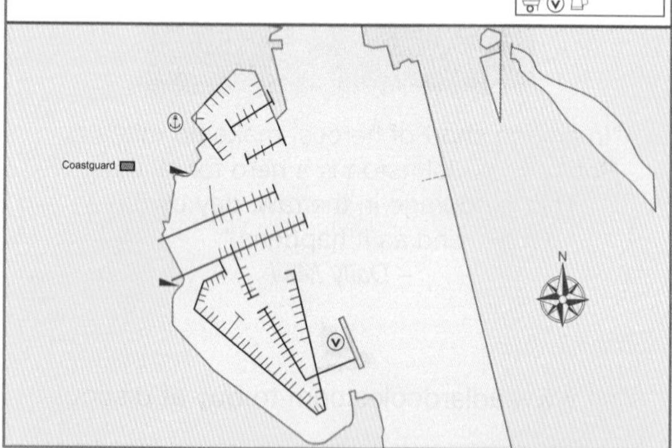

Coastguard
N

SOVEREIGN HARBOUR MARINA

Sovereign Harbour Marina
Pacific Drive, Eastbourne, East Sussex, BN23 5BJ
Tel: 01323 470099
Email: eastbourne@premiermarinas.com
www.premiermarinas.com

VHF	Ch 17
ACCESS	H24

Sovereign Harbour is Eastbourne's best kept secret, made up of five private and tranquil harbours that offer over 1000 quality berths. The marina is entered via one of two 24/7 high capacity locks that are safe and easy to navigate. The Harbour offers outstanding facilities, including a fully serviced boatyard, chandlery and self-store, plus a friendly Yacht Club and a Berth Holder's Association that offer a rich events programme. Back onshore a leisure area, 'The Waterfront', boasts a mix of bars, restaurants and cafés that overlook the water, adding to the laid-back ambiance of the marina.

FACILITIES AT A GLANCE

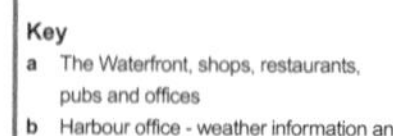

Key
- a The Waterfront, shops, restaurants, pubs and offices
- b Harbour office - weather information and visitor's information
- c Cinema
- d Retail park - supermarket and post office
- e Restaurant
- f 24 hr fuel pontoon (diesel, petrol and holding tank pump out)
- g Boatyard, boatpark, marine engineers, riggers and electricians

NB Berth numbering runs from low outer to high inner

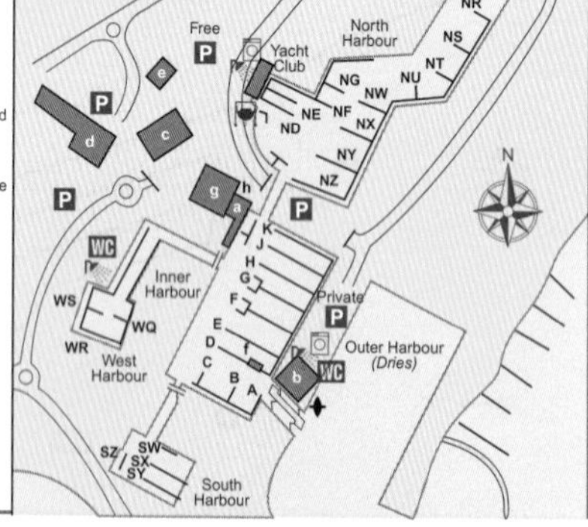

Free
Yacht Club
North Harbour
NR
NS
NT
NU
NW
NG
NE
NF
NX
ND
NY
NZ
Inner Harbour
WS
WQ
WR
West Harbour
Private
Outer Harbour
(Dries)
SZ
SW
SX
SY
South Harbour
N

HARBOUR OF RYE

Harbour of Rye
New Lydd Road, Camber, E Sussex, TN31 7QS
Tel: 01797 225225
Email: rye.harbour@environment-agency.gov.uk
www.environment-agency.gov.uk/harbourofrye

VHF	Ch 14
ACCESS	HW±2

The Strand Quay moorings are located in the centre of the historic town of Rye with all of its amenities a short walk away. The town caters for a wide variety of interests with the nearby Rye Harbour Nature Reserve, a museum, numerous antique shops and plentiful pubs, bars and restaurants. Vessels, up to a length of 15 metres, wishing to berth in the soft mud in or near the town of Rye should time their arrival at the entrance for not later than one hour after high water. Larger vessels should make prior arrangements with the Harbour Master. Fresh water, electricity, shower and toilet facilities are available.

FACILITIES AT A GLANCE

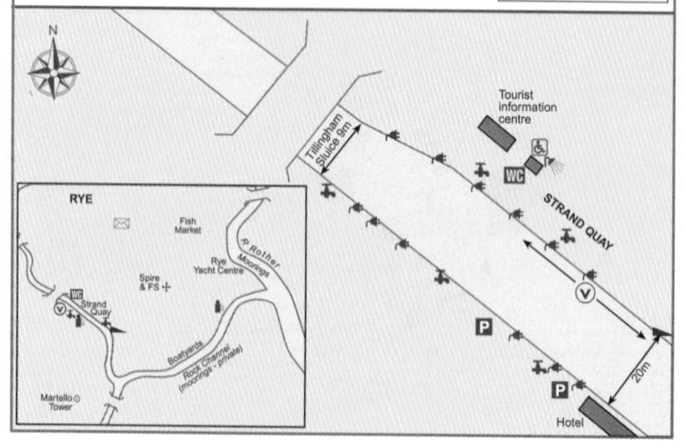

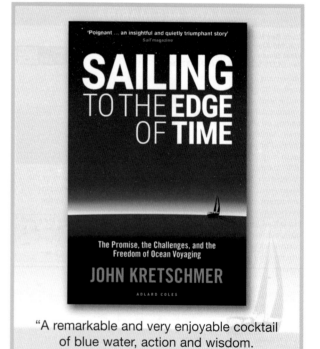

N
Tourist information centre
Tillingham Sluice 9m
STRAND QUAY
20m
Hotel
RYE
Fish Market
Rye Yacht Centre
R. Rother Moorings
Spire & FS
Strand Quay
Boatyards
Rock Channel (moorings - private)
Martello Tower

DOVER MARINA

Dover Harbour Board
Harbour House, Dover, Kent, CT17 9TF
Tel: 01304 241663 Fax: 01304 242549
Email: marina@doverport.co.uk www.doverport.co.uk/marina

VHF	Ch 80
ACCESS	H24

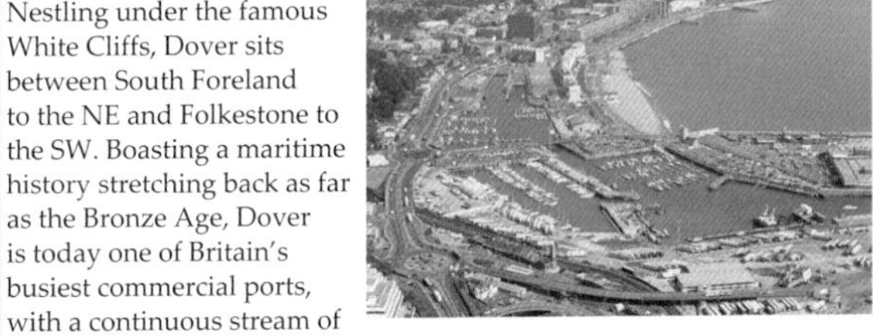

Nestling under the famous White Cliffs, Dover sits between South Foreland to the NE and Folkestone to the SW. Boasting a maritime history stretching back as far as the Bronze Age, Dover is today one of Britain's busiest commercial ports, with a continuous stream of ferries and cruise liners plying to and from their European destinations. However, over the past years the harbour has made itself more attractive to the cruising yachtsman, with the marina, set well away from the busy ferry terminal, offering three sheltered berthing options in the Tidal Harbour, Granville Dock and Wellington Dock.

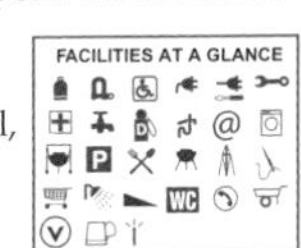
FACILITIES AT A GLANCE

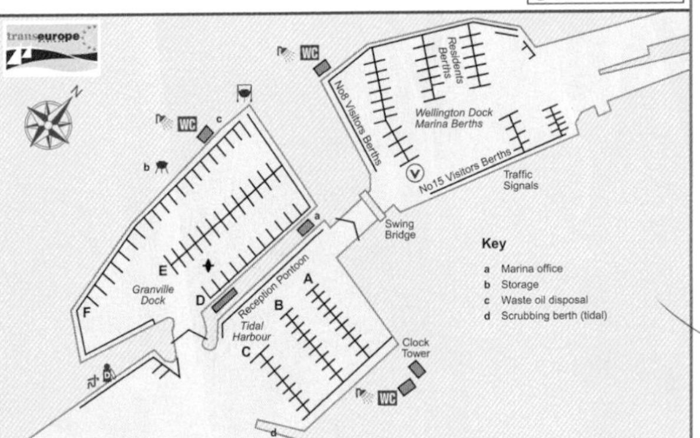
transeurope
Residents Berths
Wellington Dock Marina Berths
No6 Visitors Berths
No15 Visitors Berths
Traffic Signals
Swing Bridge
Granville Dock
Reception Pontoon
Tidal Harbour
Clock Tower
Key
a Marina office
b Storage
c Waste oil disposal
d Scrubbing berth (tidal)

ROYAL HARBOUR MARINA

Royal Harbour Marina, Ramsgate
Harbour Office, Military Road, Ramsgate, Kent, CT11 9LQ
Tel: 01843 572100
Email: portoframsgate@thanet.gov.uk
www.portoframsgate.co.uk

VHF	Ch 14, 80
ACCESS	H24

Steeped in maritime history, Ramsgate was awarded 'Royal' status in 1821 by George IV in recognition of the warm welcome he received when sailing from Ramsgate. Offering good shelter and modern facilities, including both red and white diesel, the Royal Harbour comprises an outer marina accessible H24 and an inner marina, entered approximately HW±2. Permission to enter or leave the Royal Harbour must be obtained from Port Control on channel 14 and berthing instructions can be obtained from the Dockmaster on channel 80. Full information may be found on the website.

FACILITIES AT A GLANCE

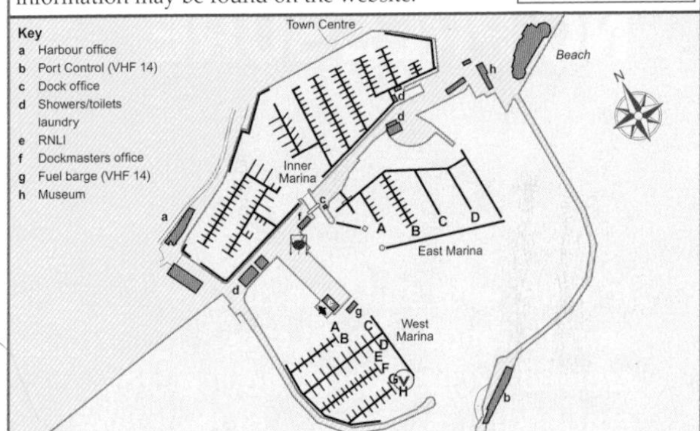
Key
a Harbour office
b Port Control (VHF 14)
c Dock office
d Showers/toilets laundry
e RNLI
f Dockmasters office
g Fuel barge (VHF 14)
h Museum
Town Centre
Beach
Inner Marina
East Marina
West Marina

EAST ENGLAND – North Foreland to Great Yarmouth

Key to Marina Plans symbols

Bottled gas	Parking
Chandler	Pub/Restaurant
Disabled facilities	Pump out
Electrical supply	Rigging service
Electrical repairs	Sail repairs
Engine repairs	Shipwright
First Aid	Shop/Supermarket
Fresh Water	Showers
Fuel - Diesel	Slipway
Fuel - Petrol	Toilets
Hardstanding/boatyard	Telephone
Internet Café	Trolleys
Laundry facilities	Visitors berths
Lift-out facilities	Wi-Fi

4

Area 4 - East England

MARINAS

Telephone Numbers, VHF Channel, Access Times

Gallions Pt Marina
020 7476 7054
Ch M, 80 HW±5

South Dock Marina
020 7252 2244
Ch M HW-2½ to +1½

Poplar Dock Marina
020 7308 9930
Ch 13 HW±1

St Katharine Haven
020 7264 5312
Ch 80 HW-2 to +1½

Chelsea Harbour
020 7225 9157
Ch 80 HW±1½

Brentford Dock Marina
020 8232 8941 HW±2½

Penton Hook Marina
01932 568681
Ch 80 H24

Windsor Marina
01753 853911
Ch 80 H24

Bray Marina
01628 623654
Ch 80 H24

Gt Yarmouth

Lowestoft

Royal Norfolk & Suffolk YC
01520 566726 Ch 14, 80 H24

Lowestoft Haven Marina
01520 580300 Ch M, 80 H24

Lowestoft Cruising Club
07913 391950, H24

Southwold

Suffolk Yacht Hbr 01473 659240 Ch 80 H24
Royal Harwich YC Marina 01473 780319 Ch 77 H24
Woolverstone Marina 01473 780206 Ch 80 H24
Fox's Marina 01473 689111 Ch 80 H24
Neptune Marina 01473 215204 Ch M, 80 H24
Ipswich Haven Marina 01473 236644 Ch M, 80 H24

Ipswich

Orford

Shotley Marina
01473 788982
Ch 80 H24

Harwich

Titchmarsh Marina
01255 672185
Ch 80 HW±5

Walton Yacht Basin
01255 675873
Ch 80 HW-¾ to +¼

Burnham Yacht Harbour 01621 782150 Ch 80 H24
Essex Marina 01702 258531 Ch 80 H24
Bridgemarsh Marina 01621 740414 Ch 80 HW±4
Fambridge Yacht Haven 01621 740370 Ch 80 H24
Fambridge Yacht Station 01621 742911 Ch 80 H24

Bradwell Marina 01621 776235 Ch M, 80 HW±4½
Blackwater Marina 01621 740264 Ch M HW±2
Tollesbury Marina 01621 869202 Ch 80 HW±2
Heybridge Basin 01621 853506 Ch 80 HW±1

Limehouse Marina
020 7308 9930
Ch 80 HW±3

R Thames

N. Foreland

Port Werburgh 01634 225107 Ch 80 HW±3

Chatham Maritime Marina
01634 899200

Gillingham Marina
01634 280022
Ch 80 HW±4½

Ramsgate

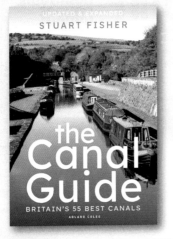

N

GILLINGHAM MARINA

Gillingham Marina
173 Pier Road, Gillingham, Kent, ME7 1UB
Tel: 01634 280022 Fax: 01634 280164
Email: berthing@gillingham-marina.co.uk
www.gillingham-marina.co.uk

VHF	Ch 80
ACCESS	HW±4.5

Gillingham Marina comprises a locked basin, accessible four and a half hours either side of high water, and a tidal basin upstream which can be entered approximately two hours either side of high water. Deep water moorings in the river cater for yachts arriving at other times.

Visiting yachts are usually accommodated in the locked basin, although it is best to contact the marina ahead of time. Lying on the south bank of the River Medway, the marina is approximately eight miles from Sheerness, at the mouth of the river, and five miles downstream of Rochester Bridge. Facilities include a well-stocked chandlery, brokerage and an extensive workshop.

FACILITIES AT A GLANCE
WC

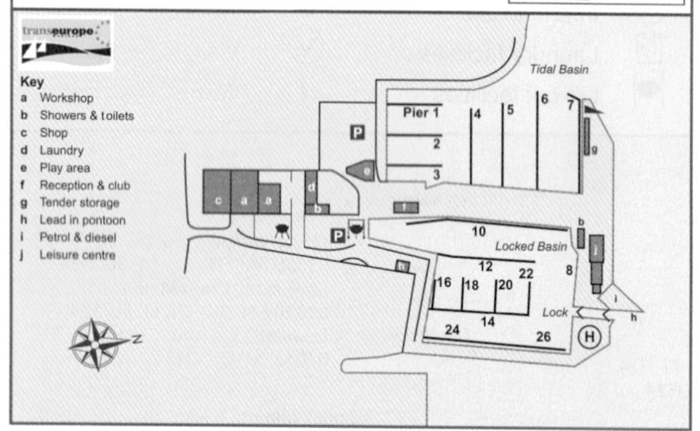
transeurope
Key
a Workshop
b Showers & toilets
c Shop
d Laundry
e Play area
f Reception & club
g Tender storage
h Lead in pontoon
i Petrol & diesel
j Leisure centre
Tidal Basin
Pier 1
2
3
4
5
6
7
P
10
Locked Basin
12
22
8
16
18
20
14
24
26
Lock
H
N

PORT WERBURGH

Port Werburgh
Vicarage Lane, Hoo, Rochester, Kent, ME3 9TW
Tel: 01634 252107 Fax: 01634 253477
Email: jillswann@wanttoliveafloat.com

VHF
Ch 80
ACCESS
HW±3

Hoo Marina has been purchased by Residential Marine Ltd and is now incorporated into Port Werburgh. The port, some eight miles upriver from Sheerness, can be approached either straight across the mud flats at HW or via the creek, which has access HW±3 for shallow draught boats. The path of the creek is marked by withies, which must be kept to port.

The port accommodates residential and leisure boats from 20–200ft and has a lifting service available for craft up to 17 tons; a dry dock is available for larger boats. There are no workshop facilities but boat owners are encouraged to work on their own boats. Security is provided by H24 CCTV coverage.

All berths are supplied with water and electricity and the onsite amenity block has showers, toilets and a laundry room. There is a grocery store adjacent and shops in Hoo village approximately half a mile distant. There is a frequent bus service to nearby Rochester.

FACILITIES AT A GLANCE
P
WC

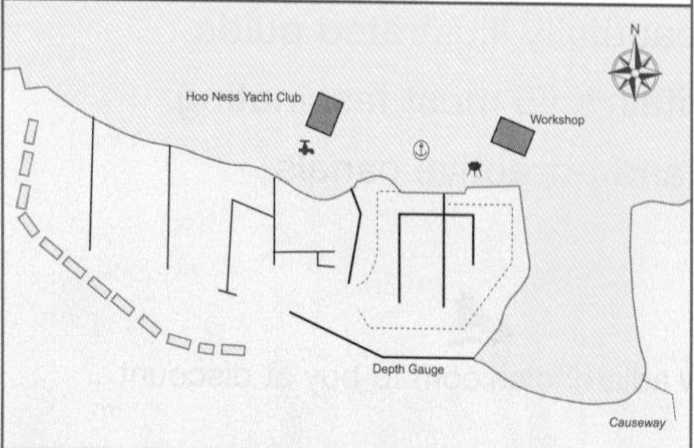
N
Hoo Ness Yacht Club
Workshop
Depth Gauge
Causeway

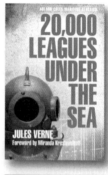

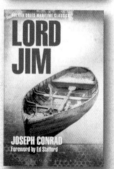

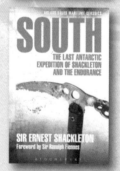

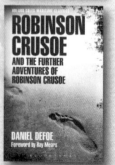

CHATHAM MARITIME MARINA

Chatham Maritime Marina, The Lock Building,
Leviathan Way, Chatham Maritime, Chatham, Medway, ME4 4LP
Tel: 01634 899200
Email: chatham@mdlmarinas.co.uk
www.chathammaritimemarina.co.uk

VHF Ch 80
ACCESS H24

Chatham Maritime Marina is situated on the banks of the River Medway in Kent, providing an ideal location from which to explore the surrounding area. There are plenty of secluded anchorages in the lower reaches of the Medway Estuary, while the river is navigable for some 13 miles from its mouth at Sheerness right up to Rochester, and even beyond for those yachts drawing less than 2m. Only 45 minutes from London by road, the marina is part of a multi-million pound leisure and retail development, accommodating 412 boats up to 24m LOA.

FACILITIES AT A GLANCE

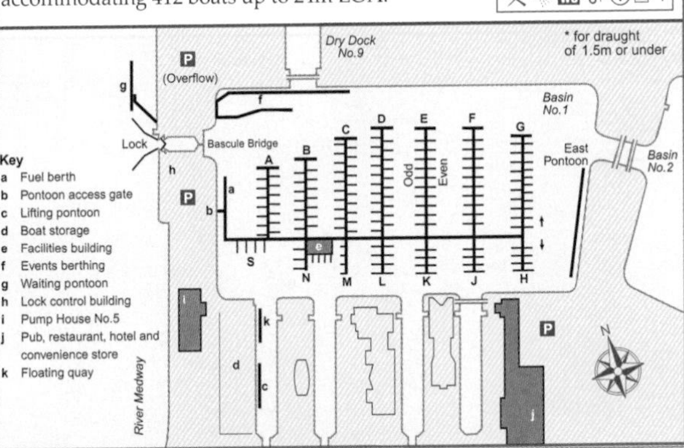
Dry Dock No.9
* for draught of 1.5m or under
(Overflow)
Basin No.1
Lock
Bascule Bridge
East Pontoon
Basin No.2
Odd
Even
Key
a Fuel berth
b Pontoon access gate
c Lifting pontoon
d Boat storage
e Facilities building
f Events berthing
g Waiting pontoon
h Lock control building
i Pump House No.5
j Pub, restaurant, hotel and convenience store
k Floating quay
River Medway

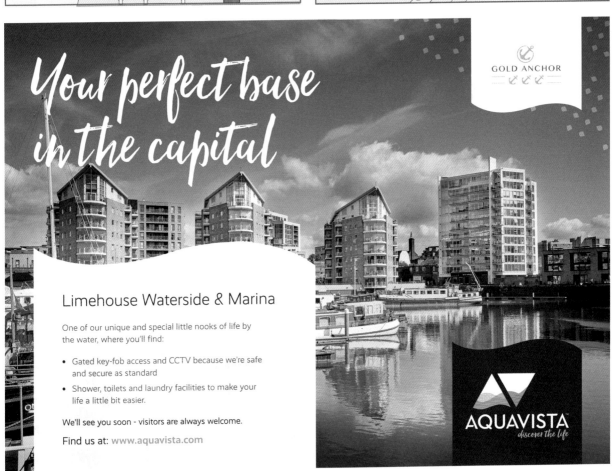

LIMEHOUSE WATERSIDE & MARINA

Limehouse Waterside & Marina
46 Goodhart Place, London, E14 8EG
Tel: 020 7308 9930
Email: limehouse@aquavista.com www.aquavista.com

VHF Ch 80
ACCESS HW±3

Limehouse Marina, situated where the canal system meets the Thames, is now considered the 'Jewel in the Crown' of the British inland waterways network. With complete access to 2,000 miles of inland waterway systems and with access to the Thames at most stages of the tide except around low water, the marina provides a superb location for river, canal and sea-going pleasure craft alike. Boasting a wide range of facilities and 137 berths, Limehouse Marina is housed in the old Regent's Canal Dock.

FACILITIES AT A GLANCE

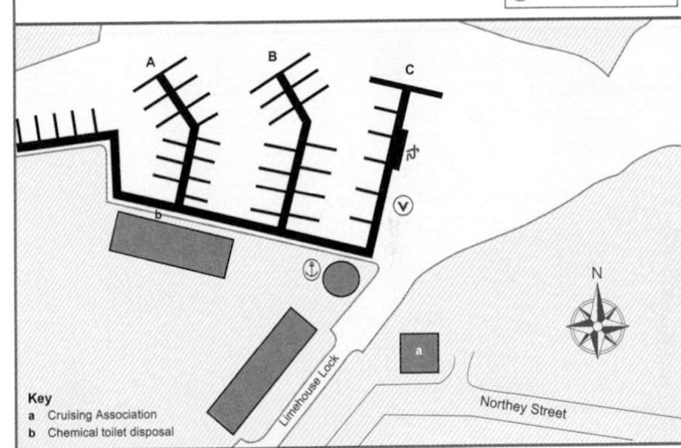
Key
a Cruising Association
b Chemical toilet disposal
Limehouse Lock
Northey Street

GALLIONS POINT MARINA

Gallions Point Marina, Gate 14, Royal Albert Basin
Woolwich Manor Way, North Woolwich
London, E16 2QY. Tel: 020 7476 7054 Fax: 020 7474 7056
Email: info@gallionspointmarina.co.uk
www.gallionspointmarina.co.uk

VHF	Ch M, 80
ACCESS	HW±5

Gallions Point Marina lies about 500 metres down-stream of the Woolwich Ferry on the north side of Gallions Reach. Accessed via a lock at the entrance to the Royal Albert Basin, the marina offers deep water pontoon berths as well as hard standing. Future plans to improve facilities include the development of a bar/restaurant, a chandlery and an RYA tuition school.

FACILITIES AT A GLANCE
WC

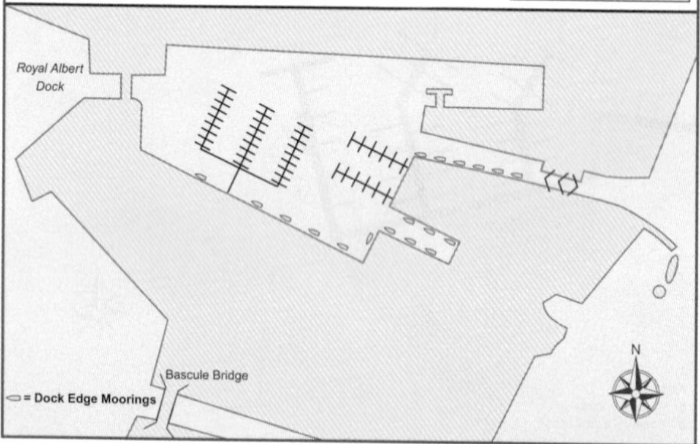
Royal Albert Dock
Bascule Bridge
= Dock Edge Moorings
N

SOUTH DOCK MARINA

South Dock Marina
Rope Street, Off Plough Way
London, SE16 7SZ
Tel: 020 7252 2244 Fax: 020 7237 3806
Email: christopher.magro@southwark.gov.uk

VHF	Ch M
ACCESS	HW-2.5 to +1.5

South Dock Marina is housed in part of the old Surrey Dock complex on the south bank of the River Thames. Its locked entrance is immediately downstream of Greenland Pier, just a few miles down river of Tower Bridge. For yachts with a 2m draught, the lock can be entered HW-2½ to HW+1½ London Bridge, although if you arrive early there is a holding pontoon on the pier. The marina can be easily identified by the conspicuous arched rooftops of Baltic Quay, a luxury waterside apartment block. Once inside this secure, 200-berth marina, you can take full advantage of all its facilities as well as enjoy a range of restaurants and bars close by or visit historic maritime Greenwich.

FACILITIES AT A GLANCE
P
WC

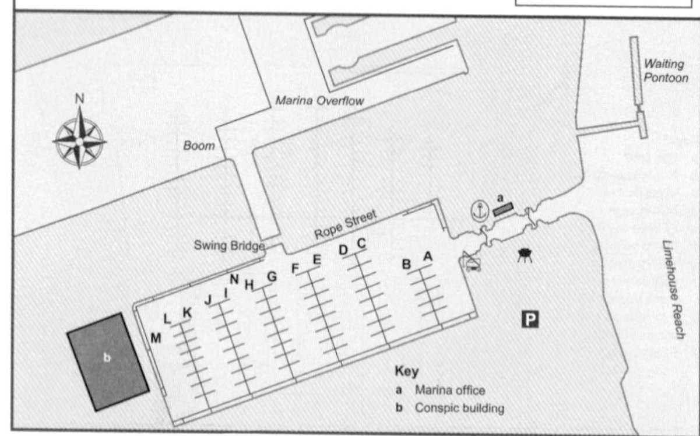
N
Marina Overflow
Boom
Waiting Pontoon
Rope Street
Swing Bridge
Limehouse Reach
A
B
C
D
E
F
G
H
N
I
J
K
L
M
a
b
P
Key
a Marina office
b Conspic building

CHELSEA HARBOUR MARINA

Chelsea Harbour Marina
Estate Managements Office
C2-3 The Chambers, London, SW10 0XF
Tel: 07770 542783 Fax: 020 7352 7868
Email: harboumaster@chelsea-harbour.co.uk

VHF	
ACCESS	HW±1.5

Chelsea Harbour is widely thought of as one of London's most significant maritime sites. It is located in the heart of SW London, therefore enjoying easy access to the amenities of Chelsea and the West End. On site is the Chelsea Harbour Design Centre, where 80 showrooms exhibit the best in British and International interior design, offering superb waterside views along with excellent cuisine in the Wyndham Grand.

The harbour lies approximately 48 miles up river from Sea Reach No 1 buoy in the Thames Estuary and is accessed via the Thames Flood Barrier in Woolwich Reach. With its basin gate operating one and a half hours either side of HW (+ 20 minutes at London Bridge), the marina welcomes visiting yachtsmen.

FACILITIES AT A GLANCE

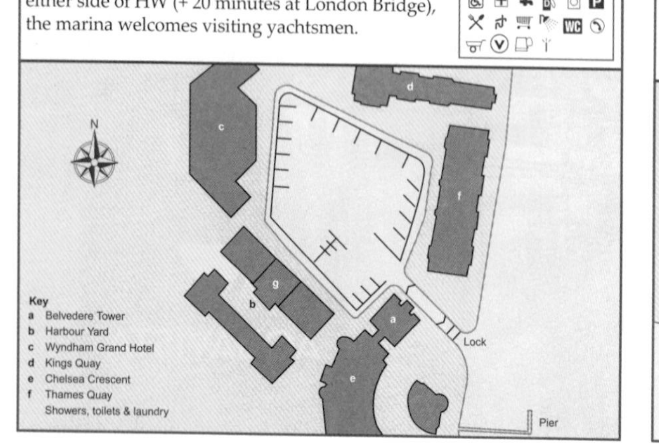
N
c
d
f
g
b
a
e
Lock
Pier
Key
a Belvedere Tower
b Harbour Yard
c Wyndham Grand Hotel
d Kings Quay
e Chelsea Crescent
f Thames Quay
Showers, toilets & laundry

ST KATHARINE DOCKS

St Katharine's Docks Marina
50 St Katharine's Way, London, E1W 1LA
Tel: 020 7264 5312
Email: marina.reception@skdocks.co.uk
www.skdocks.co.uk/marina

VHF	Ch 80
ACCESS	HW -2 to +1.5

St Katharine Docks Marina offers 185 berths equipped for boats up to 40m in three separate, secure and calm basins. The historic docks are located next to Tower Bridge.

This is a unique marina benefiting from waterside dining, boutique shops and excellent transport links to the West End. Visitors are welcomed all year round and the marina provides its own calendar of events details of which can be found on the website and social media pages.

The marina is ideally situated for visiting the Tower of London, Tower Bridge, *HMS Belfast* and the City of London all of which can be reached on foot. A short river bus service away is Greenwich and the Cutty Sark and to the west the Shard and London Eye.

FACILITIES AT A GLANCE

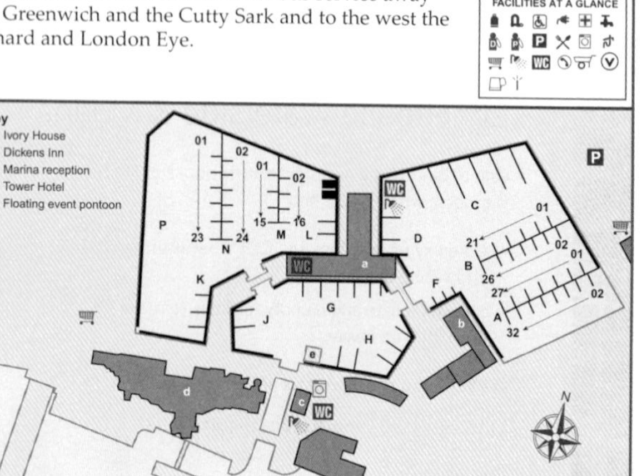
Key
a Ivory House
b Dickens Inn
c Marina reception
d Tower Hotel
e Floating event pontoon
P
WC
N

BRENTFORD DOCK MARINA

Brentford Dock Marina
2 Justin Close, Brentford, Middlesex, TW8 8QE
Tel: 020 8232 8941 Mob: 07970 143 987
E-mail: brentforddockmarina@gmail.com

VHF
ACCESS HW±2.5

Brentford Dock Marina is situated on the River Thames at the junction with the Grand Union Canal. Its hydraulic lock is accessible for up to two and a half hours either side of high water, although boats over 9.5m LOA enter on high water by prior arrangement. There is a grocery store on site. The main attractions within the area are the Royal Botanic Gardens at Kew and the Kew Bridge Steam Museum at Brentford.

FACILITIES AT A GLANCE

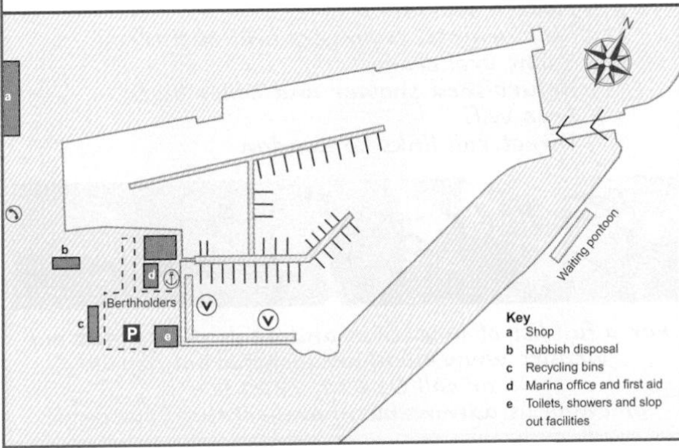
a
b
c
d
e
Berthholders
Waiting pontoon
Key
a Shop
b Rubbish disposal
c Recycling bins
d Marina office and first aid
e Toilets, showers and slop out facilities

PENTON HOOK MARINA

Penton Hook Marina
Staines Road, Chertsey, Surrey, KT16 8PY
Tel: 01932 568681
Email: pentonhook@mdlmarinas.co.uk
www.pentonhookmarina.co.uk

VHF
ACCESS H24

Penton Hook, the largest inland marina in Europe, is situated on what is considered to be one of the most attractive reaches of the River Thames; close to the vibrant town of Staines-on-Thames and about a mile downstream from Runnymede.

Providing unrestricted access to the River Thames through a deep water channel below Penton Hook Lock, the marina can accommodate ocean-going craft of up to 30m LOA and is ideally placed for a visit to Thorpe Park, reputedly one of the country's most popular family leisure attractions.

FACILITIES AT A GLANCE

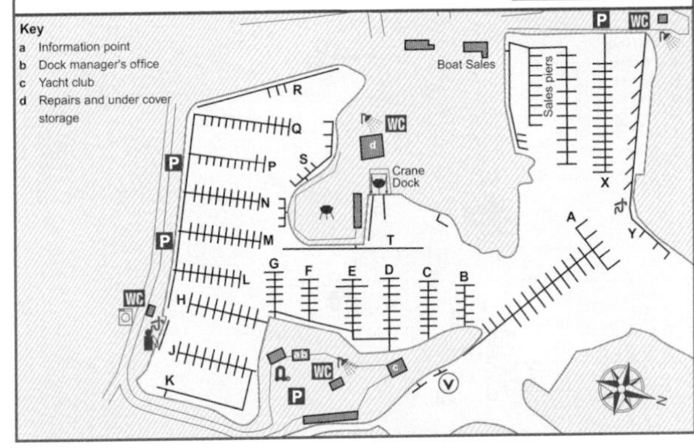
Key
a Information point
b Dock manager's office
c Yacht club
d Repairs and under cover storage
Boat Sales
Sales piers
Crane Dock
R
Q
P
N
M
L
H
J
K
S
T
G
F
E
D
C
B
A
X
Y

WINDSOR MARINA

Windsor Marina
Maidenhead Road, Windsor
Berkshire, SL4 5TZ
Tel: 01753 853911
Email: windsor@mdlmarinas.co.uk www.windsormarina.co.uk

VHF
ACCESS H24

Situated on the outskirts of Windsor town on the south bank of the River Thames, Windsor Marina enjoys a peaceful garden setting. On site is the Windsor Yacht Club, a small chandley and fuel (diesel and petrol), enabling you to fill up as and when you need.

A trip to the town of Windsor, comprising beautiful Georgian and Victorian buildings, would not be complete without a visit to Windsor Castle. With its construction inaugurated over 900 years ago by William the Conqueror, it is the oldest inhabited castle in the world and accommodates a priceless art and furniture collection.

FACILITIES AT A GLANCE

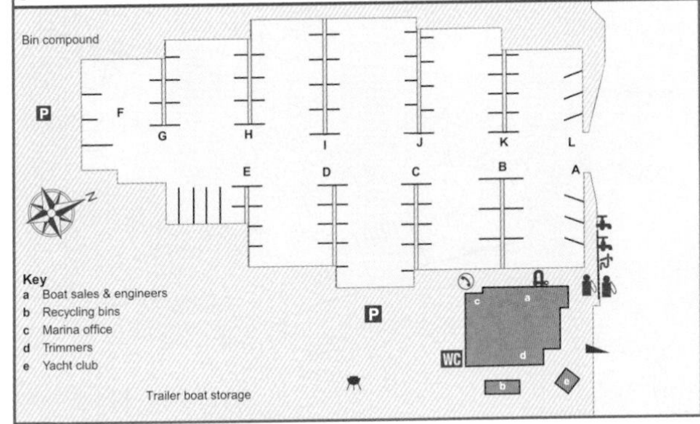
Bin compound
F
G
H
I
J
K
L
E
D
C
B
A
Key
a Boat sales & engineers
b Recycling bins
c Marina office
d Trimmers
e Yacht club
Trailer boat storage

BRAY MARINA

Bray Marina
Monkey Island Lane, Bray
Berkshire, SL6 2EB
Tel: 01628 623654
Email: bray@mdlmarinas.co.uk www.braymarina.co.uk

VHF
ACCESS H24

Bray Marina is situated in a country park setting among shady trees, providing berth holders with a tranquil mooring. From the marina there is direct access to the Thames and there are extensive well-maintained facilities available for all boat owners. The 400-berth marina boasts a Mediterranean-themed bar and restaurant, an active club, which holds social functions and boat training lessons, a small chandlery, fuel (diesel and petrol), and engineering services.

Upstream is Cliveden House, with its extensive gardens and woodlands. Cookham is home of the Queen's Swan Keeper, who can sometimes be seen in traditional costume. Further on still is Hambledon Mill; from here the river is navigable as far as Lechlade.

FACILITIES AT A GLANCE

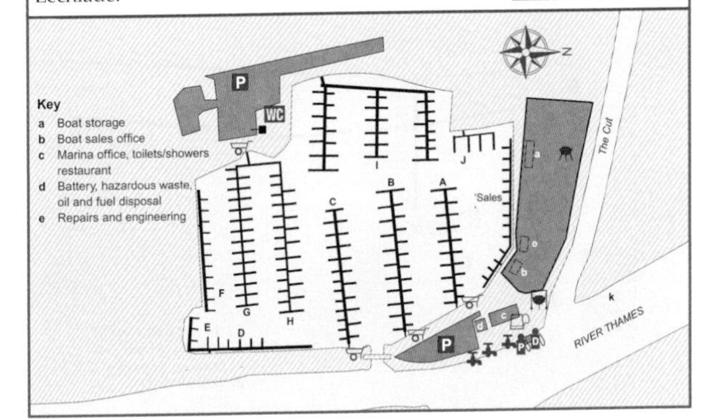
Key
a Boat storage
b Boat sales office
c Marina office, toilets/showers restaurant
d Battery, hazardous waste, oil and fuel disposal
e Repairs and engineering
The Cut
RIVER THAMES
Sales
I
J
C
B
A
F
G
H
E
D

BURNHAM YACHT HARBOUR MARINA

Burnham Yacht Harbour Marina Ltd
Burnham-on-Crouch, Essex, CM0 8BL
Tel: 01621 782150 HM: 01621 786832
Email: admin@burnhamyachtharbour.co.uk

VHF Ch 80
ACCESS H24

Burnham Yacht Harbour is situated in the pretty town of Burnham-on-Crouch with its quaint shops, elegant quayside and riverside walks. The train station, with direct train links to London Liverpool Street Station, is within walking. There is provision of all facilities you would expect to find from a modern secure marina with H24 tidal access, 350 fully serviced berths, the popular Swallowtail Restaurant and Bar, friendly staff, engineers, shipwrights, workshop, friendly staff, chandlery and yacht brokerage. The entrance is easily identified by a yellow pillar buoy with an 'X' topmark.

FACILITIES AT A GLANCE

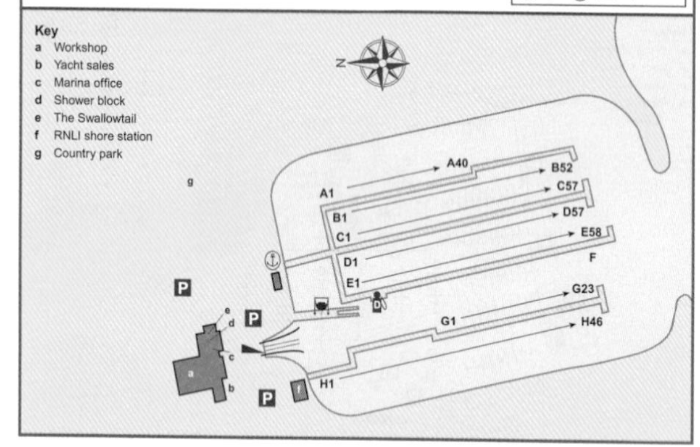
Key
a Workshop
b Yacht sales
c Marina office
d Shower block
e The Swallowtail
f RNLI shore station
g Country park
A40
B52
A1
C57
B1
D57
C1
E58
D1
F
E1
G23
G1
H46
H1

ESSEX MARINA

Essex Marina
Wallasea Island, Essex, SS4 2HF
Tel: 01702 258531 Fax: 01702 258227
Email: info@essexmarina.co.uk
www.essexmarina.co.uk

VHF Ch 80
ACCESS H24

Surrounded by beautiful countryside, Essex Marina is situated in Wallasea Bay, about half a mile up river of Burnham on Crouch. Boasting 500 deep water berths, including 50 swinging moorings, the marina can be accessed at all states of the tide. On site are a 70 ton boat hoist, a chandlery and brokerage service as well as the Essex Marina Yacht Club.

Essex Marina is the home of Boats.co.uk. There is a ferry service which runs from Easter until the end of September, taking passengers across the river 6 days a week to Burnham, where you will find numerous shops and restaurants.

FACILITIES AT A GLANCE

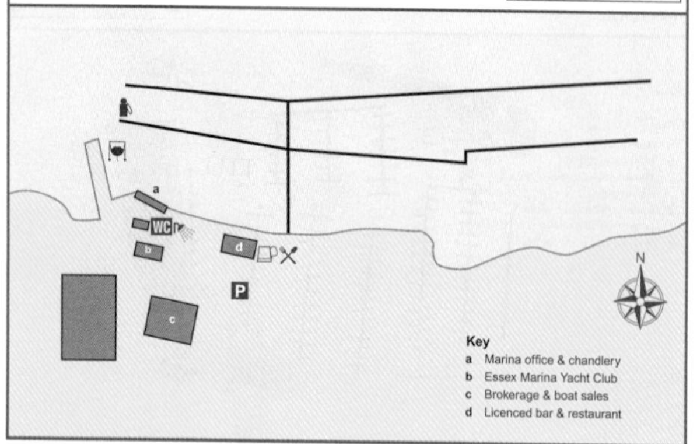
Key
a Marina office & chandlery
b Essex Marina Yacht Club
c Brokerage & boat sales
d Licenced bar & restaurant

BRIDGEMARSH MARINA

Bridgemarsh Marine
Fairholme, Bridge Marsh Lane, Althorne, Essex
Tel: 01621 740414 Mobile: 07968 696815 Fax: 01621 742216

VHF Ch 80
ACCESS HW±4

On the north side of Bridgemarsh Island, just beyond Essex Marina on the River Crouch, lies Althorne Creek. Here Bridgemarsh Marine accommodates over 100 boats berthed alongside pontoons supplied with water and electricity. A red beacon marks the entrance to the creek, with red can buoys identifying the approach channel into the marina. Accessible four hours either side of high water, the marina has an on site yard with two docks, a slipway and crane. The village of Althorne is just a short walk away, from where there are direct train services (taking approximately one hour) to London.

FACILITIES AT A GLANCE

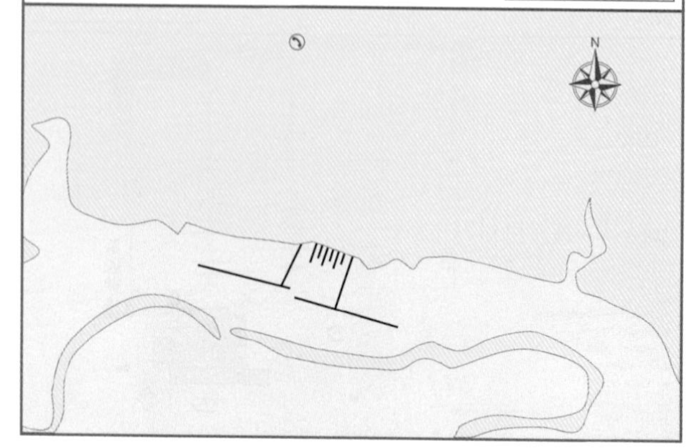
N

HEYBRIDGE BASIN

Heybridge Basin
Lock Hill, Heybridge Basin, Maldon, Essex, CM9 4RY
Tel: 07712 079764
Email: paul.hindley@waterways.org.uk
www.essexwaterways.com

VHF Ch 80
ACCESS HW±1

Towards the head of the River Blackwater and not far from Maldon, lies Heybridge Basin sea lock. It is situated at the lower end of the 14M Chelmer and Blackwater Navigation Canal and can be entered approximately 1-1.5hrs before HW for vessels drawing up to 2m. There is good holding ground in the river just outside the lock. There are in excess of 300 permanent moorings along the navigation and room for up to 20 rafting visiting vessels in the Basin, which has a range of facilities, including shower and laundry. Please book at least 24hrs in advance especially during the summer months. The manned lock is operational for tides between 0600 and 2000 during summer months (0800–1700 Apr–Oct).

FACILITIES AT A GLANCE

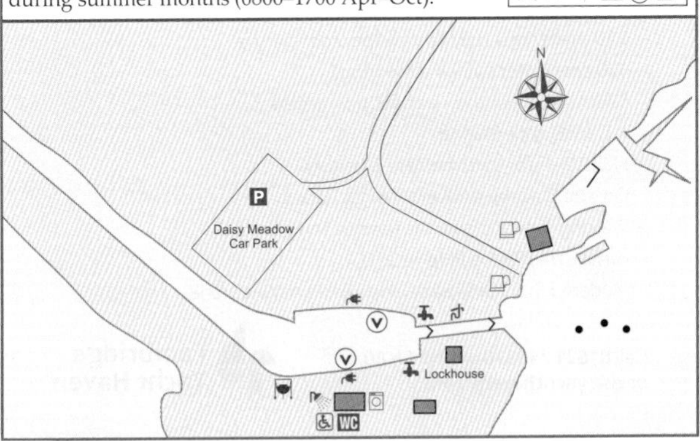
N
Daisy Meadow Car Park
Lockhouse

BRADWELL MARINA

Bradwell Marina, Port Flair Ltd, Waterside
Bradwell-on-Sea, Essex, CM0 7RB
Tel: 01621 776235 Fax: 01621 776393
Email: info@bradwellmarina.com
www.bradwellmarina.com

VHF Ch M, 80
ACCESS HW±4.5

Opened in 1984, Bradwell is a privately-owned marina situated in the mouth of the River Blackwater, serving as a convenient base from which to explore the Essex coastline or as a departure point for cruising further afield to Holland and Belgium.

The yacht basin can be accessed four and a half hours either side of HW and offers plenty of protection from all wind directions. With a total of 350 fully serviced berths, generous space has been allocated for manoeuvring between pontoons. Overlooking the marina is Bradwell Club House, incorporating a bar, restaurant, launderette and ablution facilities.

FACILITIES AT A GLANCE

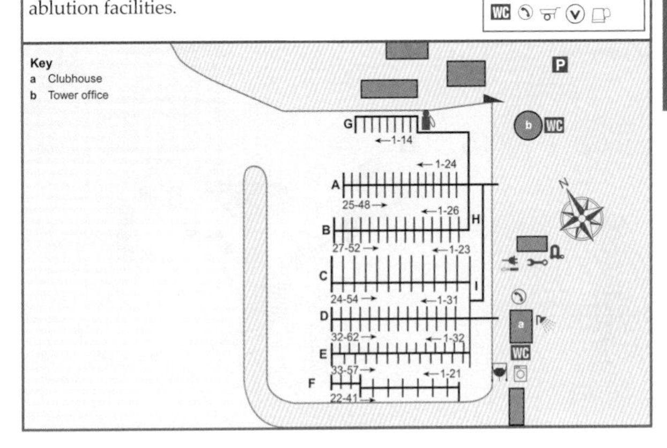
Key
a Clubhouse
b Tower office
G
1-14
1-24
A
25-48
1-26
B
H
27-52
1-23
C
24-54
1-31
I
D
32-62
1-32
E
33-57
1-21
F
22-41

FAMBRIDGE YACHT STATION

Fambridge Yacht Station
Church Road, North Fambridge, Essex, CM3 6LU
Tel: 01621 742911 www.yachthavens.com
Email: fambridge@yachthavens.com

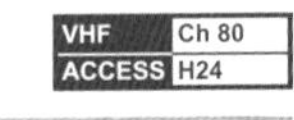
VHF Ch 80
ACCESS H24

The Yacht Station is located just under a mile downstream of Fambridge Yacht Haven and is set within the sheltered River Crouch directly between the rural villages of North and South Fambridge. Home to the North Fambridge Yacht Club, the Yacht Station has a 120m visitor pontoon providing deep water berthing alongside and foot access to mud berths and the North Fambridge. There are also 120 deep water swinging moorings in 4 straight E/W trots just off the visitor pontoon. A launch service operates 7 days a week during the summer, call in advance for times. As with the marina, excellent repair facilities can be found ashore for vessels with a maximum weight of 25t.

FACILITIES AT A GLANCE

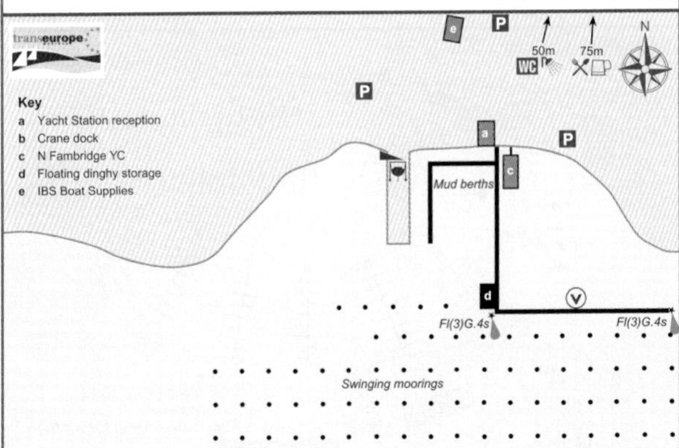
transeurope
Key
a Yacht Station reception
b Crane dock
c N Fambridge YC
d Floating dinghy storage
e IBS Boat Supplies
50m
75m
N
Mud berths
Fl(3)G.4s
Fl(3)G.4s
Swinging moorings

FAMBRIDGE YACHT HAVEN

Fambridge Yacht Haven
Church Road, North Fambridge, Essex, CM3 6LU
Tel: 01621 740370 www.yachthavens.com
Email: fambridge@yachthavens.com

VHF
Ch 80
ACCESS
H24

The Haven is split over two sites in the village of North Fambridge. The marina is located just under a mile upstream of its 'Yacht Station' facility with access via Stow Creek, which branches N off the R Crouch. The channel is straight and clearly marked to the ent of the Yacht Haven. Home to the West Wick Yacht Club, the marina has 220 berths to accommodate vessels up to 20m LOA. North Fambridge features the 500 year-old Ferry Boat Inn, a favourite haunt with the sailing fraternity. Burnham-on-Crouch is six miles down river, while the Essex and Kent coasts are within easy sailing distance. Excellent repair facilities can be found ashore together with undercover storage for vessels up to 19m LOA with a maximum weight of 40 tons.

FACILITIES AT A GLANCE

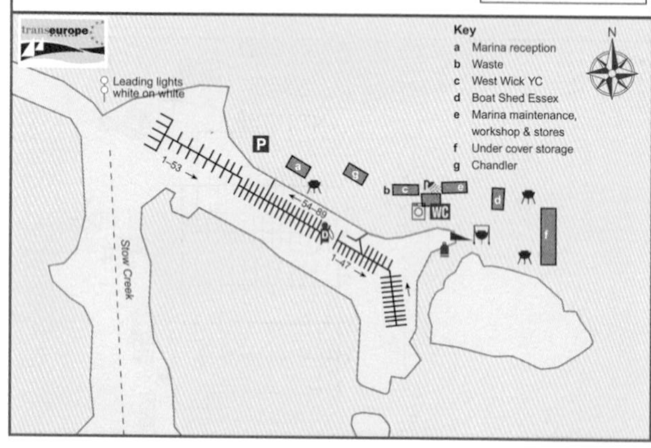

transeurope
Key
a Marina reception
b Waste
c West Wick YC
d Boat Shed Essex
e Marina maintenance, workshop & stores
f Under cover storage
g Chandler
N
Leading lights white on white
Stow Creek
1–53
54–89
1–47

BLACKWATER MARINA

Blackwater Marina
Marine Parade, Maylandsea, Essex
Tel: 01621 740264
Email: info@blackwater-marina.co.uk

VHF	Ch M
ACCESS	HW±2

Blackwater Marina is a place where families in day boats mix with Smack owners and yacht crews; here seals, avocets and porpoises roam beneath the big, sheltering East Coast skies and here the area's rich heritage of working Thames Barges and Smacks remains part of daily life today.

But it isn't just classic sailing boats that thrive on the Blackwater. An eclectic mix of motor cruisers, open boats and modern yachts enjoy the advantages of a marina sheltered by its natural habitat, where the absence of harbour walls allows uninterrupted views of some of Britain's rarest wildlife and where the 21st century shoreside facilities are looked after by experienced professionals, who are often found sailing on their days off.

FACILITIES AT A GLANCE

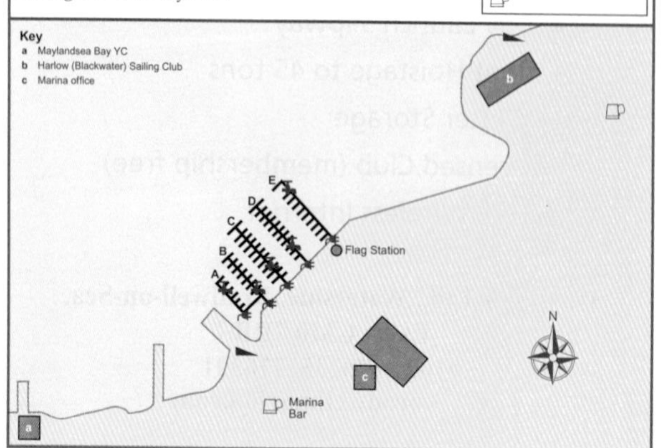

Key
a Maylandsea Bay YC
b Harlow (Blackwater) Sailing Club
c Marina office
Flag Station
Marina Bar
N

TOLLESBURY MARINA

Tollesbury Marina
The Yacht Harbour, Tollesbury, Essex, CM9 8SE
Tel: 01621 869202 Fax: 01621 868489
email: harbourmaster@tollesburymarina.com

VHF	Ch 80
ACCESS	HW±2

Tollesbury Marina lies at the mouth of the River Blackwater in the heart of the Essex countryside. Within easy access from London and the Home Counties, it has been designed as a leisure centre for the whole family, with on-site activities comprising tennis courts and a covered heated swimming pool as well as a convivial bar and restaurant. Accommodating over 240 boats, the marina can be accessed two hours either side of HW and is ideally situated for those wishing to explore the River Crouch to the south and the Rivers Colne, Orwell and Deben to the north.

FACILITIES AT A GLANCE

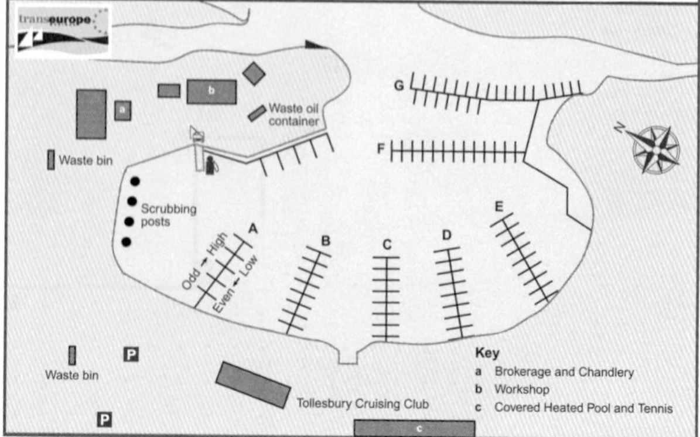

transeurope
Waste oil container
Waste bin
Scrubbing posts
Odd → High
Even ← Low
Waste bin
Tollesbury Cruising Club
Key
a Brokerage and Chandlery
b Workshop
c Covered Heated Pool and Tennis

TITCHMARSH MARINA

Titchmarsh Marina Ltd
Coles Lane, Walton on the Naze, Essex, CO14 8SL
Tel: 01255 672185 Fax: 01255 851901
Email: info@titchmarshmarina.co.uk
www.titchmarshmarina.co.uk

VHF	Ch 80
ACCESS	HW±5

Titchmarsh Marina sits on the south side of The Twizzle in the heart of the Walton Backwaters. As the area is designated a 'wetland of international importance', the marina has been designed and developed to function as a natural harbour. The 420 berths are well-sheltered by the high-grassed clay banks, offering good protection in all conditions. The marina entrance has a depth of 1.3m at LWS but once inside the basin this increases to around 2m; there is a tide gauge at the fuel berth. Among the excellent facilities onsite are the well-stocked chandlery and the Harbour Lights restaurant and bar serving food daily.

FACILITIES AT A GLANCE

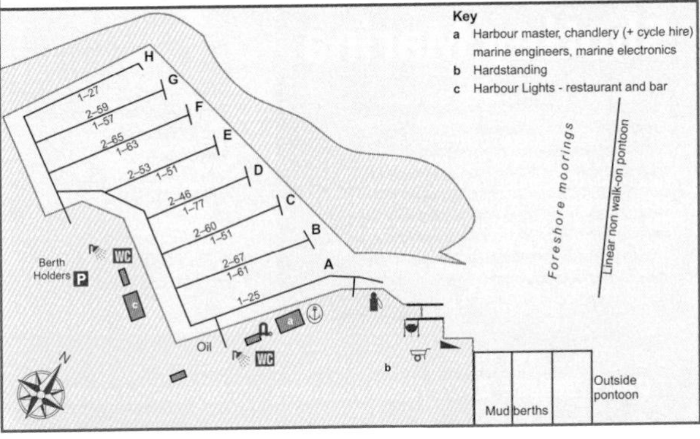
Key
a Harbour master, chandlery (+ cycle hire) marine engineers, marine electronics
b Hardstanding
c Harbour Lights - restaurant and bar
Foreshore moorings
Linear non walk-on pontoon
Berth Holders
Oil
Outside pontoon
Mud berths

TITCHMARSH marina
Friendly service in the beautiful Walton Backwaters
www.titchmarshmarina.co.uk
Tel: 01255 672 185 VHF: Channel 80
Email: info@titchmarshmarina.co.uk
Berthing Available - Various Options & Rates
Winter - Special Rates Ashore or Afloat
Storage Ashore - Long/ Short Term
Full Marina Facilities
The Yacht Harbour Association Ltd
Walton-on-the-Naze, Essex CO14 8SL
BRITISH MARINE LEADING THE INDUSTRY

WALTON YACHT BASIN

Walton and Frinton Yacht Trust
Mill Lane, Walton on the Naze, CO14 8PF
Managed by Bedwell & Co Tel: 01255 675873
Mobile: 07957 848031

VHF	Ch 80
ACCESS	HW-0.75,HW+0.25

Walton Yacht Basin lies at the head of Walton Creek, an area made famous in Arthur Ransome's *Swallows & Amazons* and *Secret Waters*. The creek can only be navigated HW±2, although yachts heading for the Yacht Basin should arrive on a rising tide as the entrance gate is kept shut once the tide turns in order to retain the water inside. Before entering the gate, moor up against the Club Quay to enquire about berthing availability.

A short walk away is the popular seaside town of Walton, full of shops, pubs and restaurants. Its focal point is the pier which, overlooking superb sandy beaches, offers various attractions. Slightly further out of town, the Naze affords pleasant coastal walks with striking panoramic views.

FACILITIES AT A GLANCE

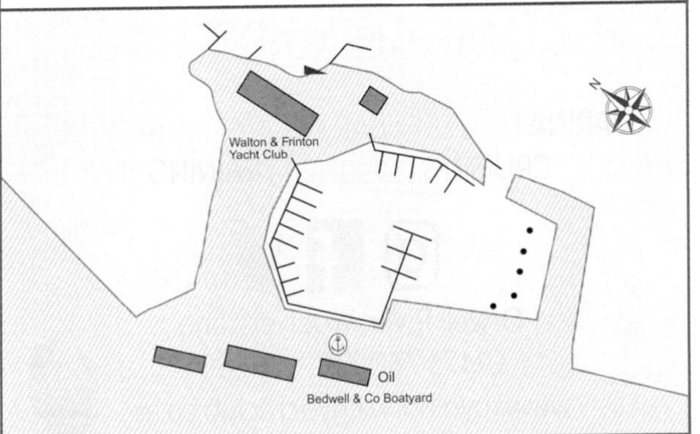
Walton & Frinton Yacht Club
Oil
Bedwell & Co Boatyard

SUFFOLK YACHT HARBOUR

Suffolk Yacht Harbour Ltd
Levington, Ipswich, Suffolk, IP10 0LN
Tel: 01473 659240 Fax: 01473 659632
Email: enquiries@syharbour.co.uk
www.syharbour.co.uk

VHF	Ch 80
ACCESS	H24

A friendly, independently-run marina on the East Coast of England, Suffolk Yacht Harbour enjoys a beautiful rural setting on the River Orwell, yet is within easy access of Ipswich, Woodbridge and Felixstowe. With approximately 550 berths, the marina offers extensive facilities while the Haven Ports Yacht Club provides a bar and restaurant.

FACILITIES AT A GLANCE

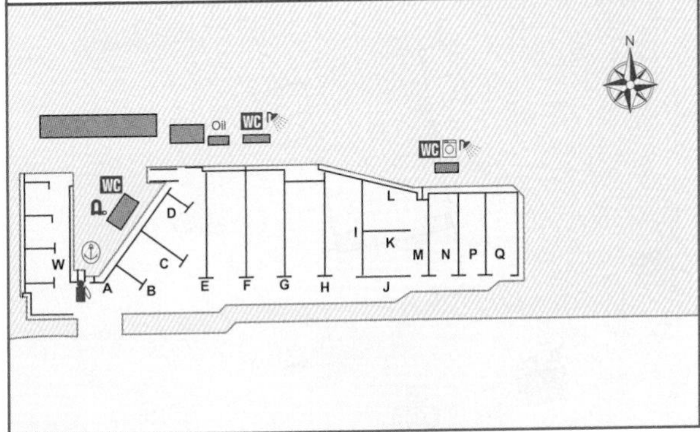
Oil
A B C D E F G H I J K L M N P Q W

SHOTLEY MARINA

Shotley Marina Ltd
Shotley Gate, Ipswich,Suffolk, IP9 1QJ
Tel: 01473 788982 Fax: 01473 788868
Email: sales@shotleymarina.co.uk
www.shotleymarina.co.uk

VHF	Ch 80
ACCESS	H24

Based in the well protected Harwich Harbour where the River Stour joins the River Orwell, Shotley Marina is only eight miles from the county town of Ipswich. Entered via a lock at all states of the tide, its first class facilities include extensive boat repair and maintenance services as well as a well-stocked chandlery and on site bar and restaurant. The marina is strategically placed for sailing up the Stour to Manningtree, up the Orwell to Pin Mill or exploring the Rivers Deben, Crouch and Blackwater as well as the Walton Backwaters.

FACILITIES AT A GLANCE

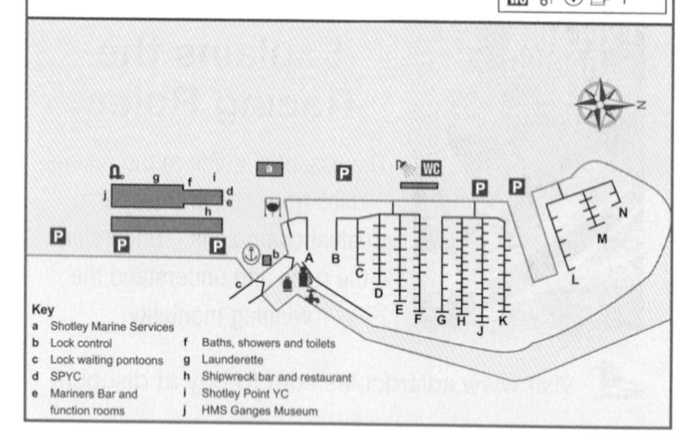
Key
a Shotley Marine Services
b Lock control
c Lock waiting pontoons
d SPYC
e Mariners Bar and function rooms
f Baths, showers and toilets
g Launderette
h Shipwreck bar and restaurant
i Shotley Point YC
j HMS Ganges Museum

"The East Coast's Premier Marina..."
shotleymarina.co.uk
Shotley Marina
The marina is situated where the rivers Stour and Orwell meet in Harwich Harbour. Access to this sheltered marina is via a hydraulically operated lock, manned continually throughout the year, 24 hours a day. The lockmaster listens on VHF CH80. The entrance to the lock is floodlit at night and the outer end of the access channel has lit piles. An "Inogon" guidance light system assists on entering the lock.
350 Ultra Modern Berths
Safe & Secure Mooring
24 Hour Access
Specialist Boat Repairs
Public House & Restaurant
Water, Electricity, Modern Toilets, Showers, Laundry and Disabled Facilities
Ships Stores
WiFi Internet Access
Shotley Marina Ltd. Shotley Gate, Ipswich, Suffolk IP9 1QJ
T: 01473 788982 F: 01473 788868 E: sales@shotleymarina.co.uk

ROYAL HARWICH YACHT CLUB MARINA

Royal Harwich Yacht Club Marina
Marina Road, Woolverstone, Suffolk, IP9 1AT
Tel: 01473 780319 Fax: 01473 780919 Berths: 07742 145994
www.royalharwichyachtclub.co.uk
Email: office.manager@royalharwich.co.uk

VHF	Ch 77
ACCESS	H24

This 54 berth marina is ideally situated at a mid point on the Orwell between Levington and Ipswich. The facility is owned and run by the Royal Harwich Yacht Club and enjoys a full catering and bar service in the Clubhouse. The marina benefits from full tidal access, and can accommodate yachts up to 14.5m on the hammerhead. Within the immediate surrounds, there are boat repair services, and a well stocked chandlery. The marina is situated a mile's walk from the world famous Pin Mill and is a favoured destination with visitors from Holland, Belgium and Germany. The marina welcomes racing yachts and cruisers, and is able to accommodate multiple bookings.

FACILITIES AT A GLANCE

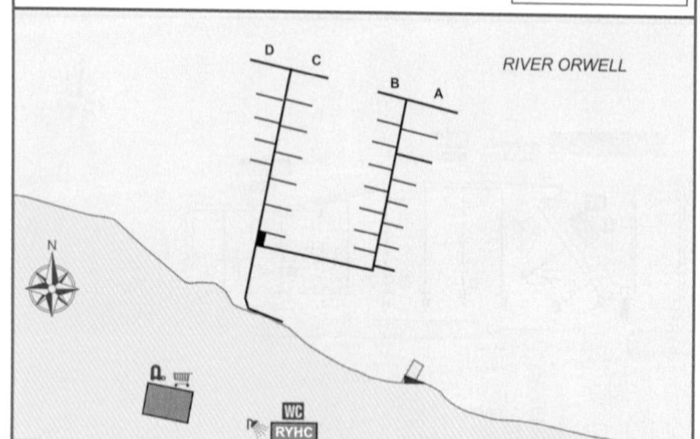
RIVER ORWELL
RYHC

ROYAL HARWICH YACHT CLUB

A FANTASTIC DESTINATION FOR EAST COAST CRUISING

COME AND VISIT US

MARINA | GALLEY | BAR | GREAT CLUBHOUSE
CRUISING | RACING | TRAINING

175 YEARS
1843-2018

Orwell River, Woolverstone
01473 780 319 ~ VHF80
www.royalharwichyachtclub.co.uk

SAILING
WINNER

FOX'S MARINA

Fox's Marina & Boatyard
The Strand, Ipswich, Suffolk, IP2 8SA
Tel: 01473 689111
Email: foxs@foxsmarina.com www.foxsmarina.com

VHF Ch 80
ACCESS H24

Located on the picturesque River Orwell, Fox's provides good shelter in all conditions and access at all states of tide with 100 pontoon berths and ashore storage for 200 vessels. A 70T hoist is able to handle boats up to 80ft in length.

Fox's Marina & Boatyard offers a full range of in-house services and, with 10,000 sq ft of heated workshop space, are specialists in repairs and refits of sailing/motor yachts and commercial craft. Specific services include coppercoat and osmosis treatment, specialist GRP and gelcoat repairs, spray painting and varnishing, and custom stainless fabrication. Also on-site, Fox's Chandlery and Marine Store, is the largest stockist of marine chandlery and equipment, sailing, leisure and country clothing in East Anglia.

FACILITIES AT A GLANCE

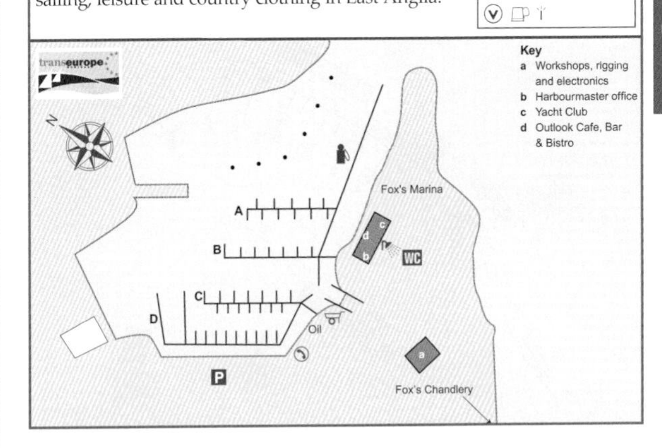
transeurope
Key
a Workshops, rigging and electronics
b Harbourmaster office
c Yacht Club
d Outlook Cafe, Bar & Bistro
Fox's Marina
A
B
C
D
WC
Oil
P
Fox's Chandlery

WOOLVERSTONE MARINA

Woolverstone Marina and Lodge Park
Woolverstone, Ipswich, Suffolk, IP9 1AS
Tel: 01473 780206
Email: woolverstone@mdlmarinas.co.uk
www.woolverstonemarina.co.uk

VHF Ch 80
ACCESS H24

Woolverstone Marina is set in 22 acres of glorious parkland on the picturesque River Orwell. Within easy reach of the sea and a multitude of scenic destinations, this is a great base to start cruising. Walton Backwaters and the River Deben are only a short distance away. If you prefer longer distance cruising then Belgium and Holland are directly across the North Sea.

Besides boat repair services and ablution facilities, the marina also incorporates a lodge park with 15 luxurious lodges, yacht brokerage and an on-site restaurant/bar, which overlooks the river.

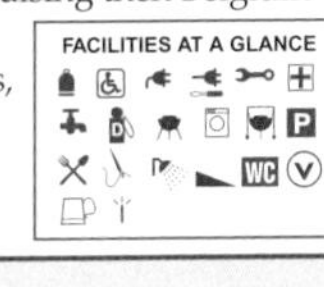
FACILITIES AT A GLANCE

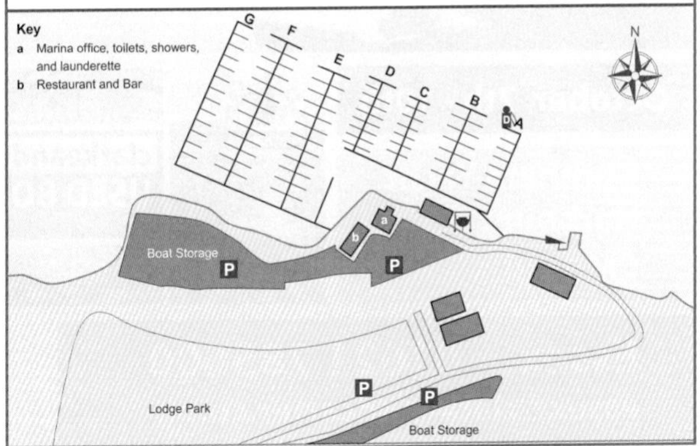
Key
a Marina office, toilets, showers, and launderette
b Restaurant and Bar
G
F
E
D
C
B
N
Boat Storage
P
Lodge Park
Boat Storage

4

NEPTUNE MARINA

Neptune Marina Ltd
Neptune Quay, Ipswich, IP4 1QJ
Tel: 01473 215204
Email: enquiries@neptune-marina.com

VHF	Ch M, 80
ACCESS	H±2.5

Neptune Marina is situated at Neptune Quay on the historic waterfront, and ever-increasing shoreside developments. This 26-acre dock is accessible through a H24 lock gate, with a waiting pontoon outside. Onsite facilities include boatyard and lift-out facilities plus superfast wifi for boat owners.

The Neptune Marina building occupies an imposing position in the NE corner of the dock with quality coffee shop and associated retail units. There are a number of excellent restaurants along the quayside and adjacent to the marina.

The modern town centre catering for all needs is just a 10-minute walk away.

FACILITIES AT A GLANCE

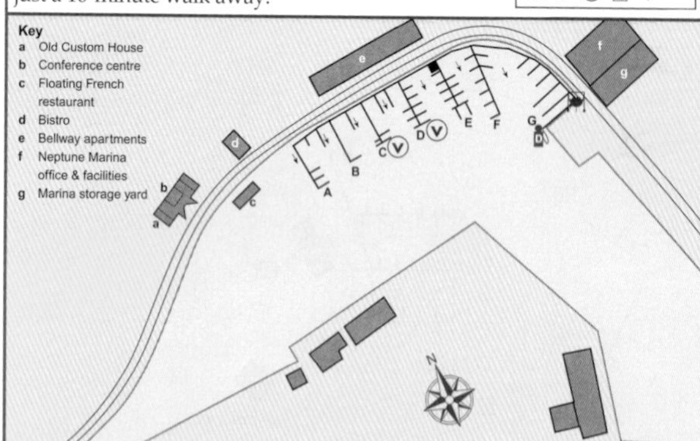
Key
a Old Custom House
b Conference centre
c Floating French restaurant
d Bistro
e Bellway apartments
f Neptune Marina office & facilities
g Marina storage yard
A
B
C
D
E
F
G
N

IPSWICH HAVEN MARINA

Ipswich Haven Marina
Associated British Ports
New Cut East, Ipswich, Suffolk, IP3 0EA
Tel: 01473 236644 Fax: 01473 236645
Email: ipswichhaven@abports.co.uk

VHF	Ch M, 80
ACCESS	H24

Lying at the heart of Ipswich, the Haven Marina enjoys close proximity to all the bustling shopping centres, restaurants, cinemas and museums that this County Town of Suffolk has to offer. The main railway station is only a 10-minute walk away, where there are regular connections to London, Cambridge and Norwich, all taking just over an hour to get to.

Within easy reach of Holland, Belgium and Germany, East Anglia is proving an increasingly popular cruising ground. The River Orwell, displaying breathtaking scenery, was voted one of the most beautiful rivers in Britain by the RYA.

FACILITIES AT A GLANCE

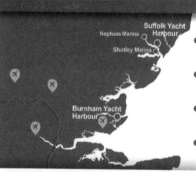

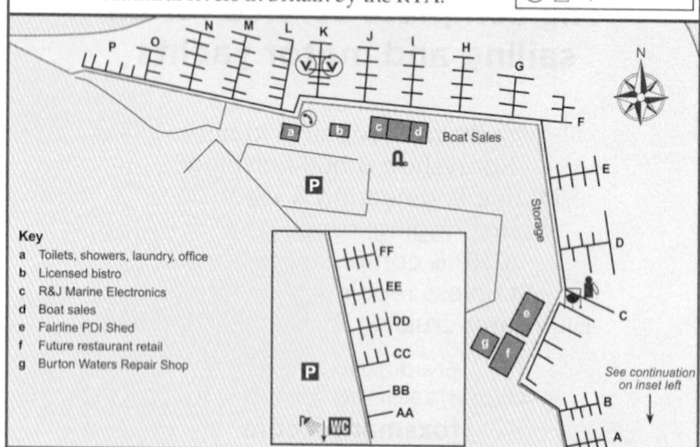
Key
a Toilets, showers, laundry, office
b Licensed bistro
c R&J Marine Electronics
d Boat sales
e Fairline PDI Shed
f Future restaurant retail
g Burton Waters Repair Shop
Boat Sales
Storage
See continuation on inset left
P O N M L K J I H G F E D C B A
FF EE DD CC BB AA
N

LOWESTOFT HAVEN MARINA

Lowestoft Haven Marina
School Road, Lowestoft, Suffolk, NR33 9NB
Tel: 01502 580300
Email: lowestofhaven@abports.co.uk
www.lowestofthavenmarina.co.uk

VHF	Ch M, 80
ACCESS	H24

Lowestoft Haven Marina is based on Lake Lothing with easy access to both the open sea and the Norfolk Broads. The town centres of both Lowestoft and Oulton Broad are within a short distance of the marina.

The marina's 140 berths can accommodate vessels from 7–20m. Offering a full range of modern facilities the marina welcomes all visitors.

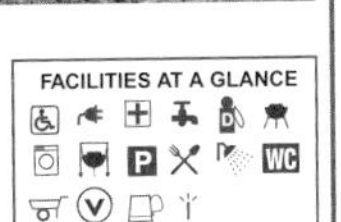
FACILITIES AT A GLANCE

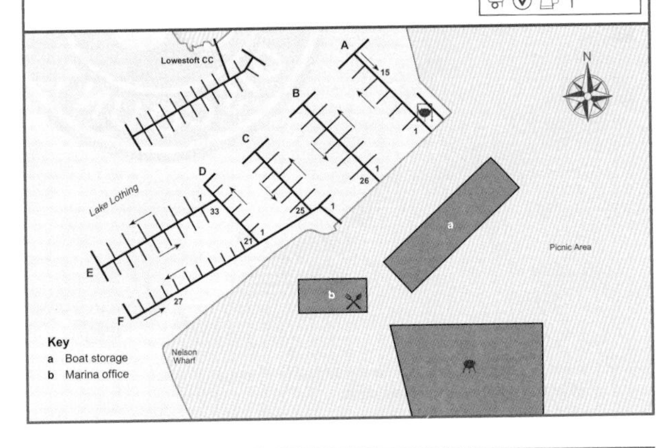
Lowestoft CC
Lake Lothing
Picnic Area
Nelson Wharf
Key
a Boat storage
b Marina office

ROYAL NORFOLK & SUFFOLK YACHT CLUB

Royal Norfolk and Suffolk Yacht Club
Royal Plain, Lowestoft, Suffolk, NR33 0AQ
Tel: 01502 566726
Email: admin@rnsyc.org.uk www.rnsyc.net

VHF	Ch 14, 80
ACCESS	H24

With its entrance at the inner end of the South Pier, opposite the Trawl Basin on the north bank, the Royal Norfolk and Suffolk Yacht Club marina occupies a sheltered position in Lowestoft Harbour. Lowestoft has always been an appealing destination to yachtsmen due to the fact that it can be accessed at any state of the tide, 24 hours a day. Note, however, that conditions just outside the entrance can get pretty lively when the wind is against tide. The clubhouse is enclosed in an impressive Grade 2 listed building overlooking the marina and its facilities include a bar and restaurant as well as a formal dining room.

FACILITIES AT A GLANCE

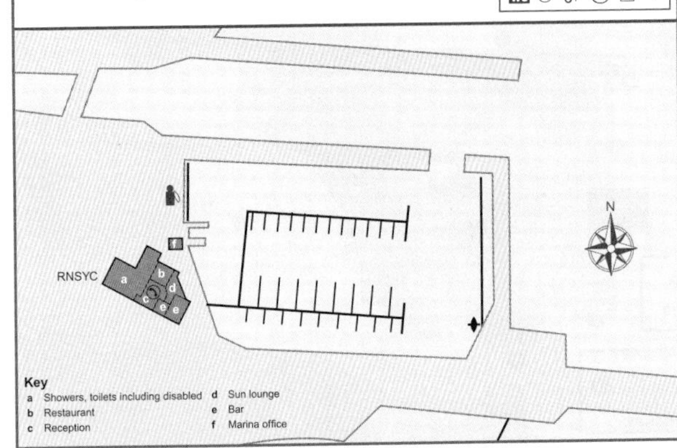
RNSYC
Key
a Showers, toilets including disabled
b Restaurant
c Reception
d Sun lounge
e Bar
f Marina office

LOWESTOFT CRUISING CLUB

Lowestoft Cruising Club
Off Harbour Road, Oulton Broad, Lowestoft, Suffolk, NR32 3LY
Tel: 07810 522515
www.lowestoftcruisingclub.co.uk

VHF	
ACCESS	H24

Lowestoft Cruising Club welcomes visitors and can offer a friendly atmosphere, some of the finest moorings and at very competitive rates. Whatever the weather, these moorings provide a calm, safe haven for visiting yachts and with the Mutford lock only 250 metres away, easy access onto the Norfolk and Suffolk Broads. Facilities include electricity and water, plus excellent showers, toilets and secure car parking. These moorings are the nearest ones to the railway stations (to Norwich and Ipswich), bus routes, shops, banks, pubs and restaurants in Oulton Broad.

FACILITIES AT A GLANCE

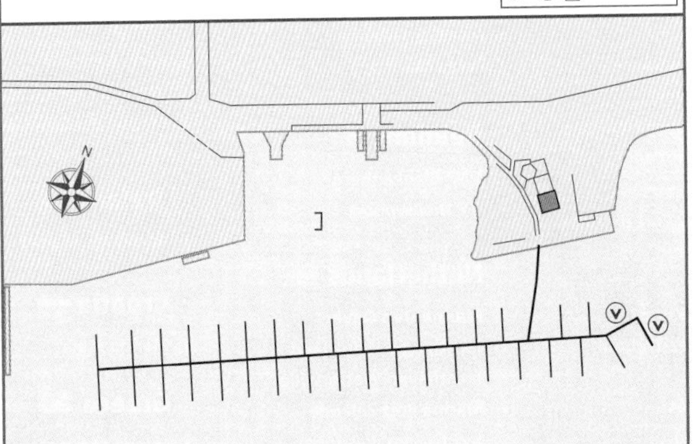

NORTH EAST ENGLAND - Great Yarmouth to Berwick-upon-Tweed

Key to Marina Plans symbols

- Bottled gas
- Chandler
- Disabled facilities
- Electrical supply
- Electrical repairs
- Engine repairs
- First Aid
- Fresh Water
- Fuel - Diesel
- Fuel - Petrol
- Hardstanding/boatyard
- Internet Café
- Laundry facilities
- Lift-out facilities
- Parking
- Pub/Restaurant
- Pump out
- Rigging service
- Sail repairs
- Shipwright
- Shop/Supermarket
- Showers
- Slipway
- Toilets
- Telephone
- Trolleys
- Visitors berths
- Wi-Fi

5

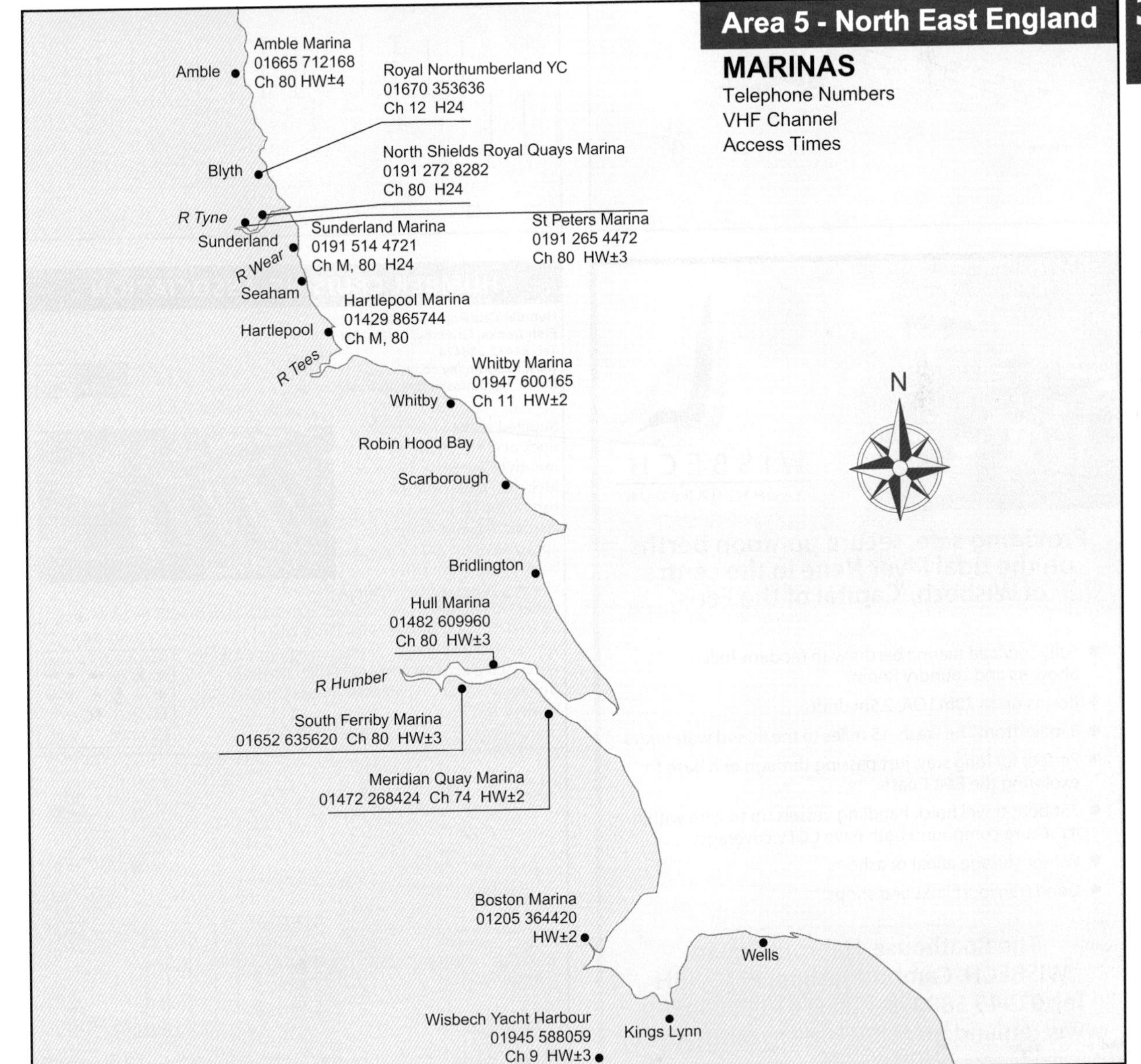
Area 5 - North East England
MARINAS
Telephone Numbers
VHF Channel
Access Times
Amble
Amble Marina
01665 712168
Ch 80 HW±4
Royal Northumberland YC
01670 353636
Ch 12 H24
Blyth
North Shields Royal Quays Marina
0191 272 8282
Ch 80 H24
R Tyne
St Peters Marina
0191 265 4472
Ch 80 HW±3
Sunderland
Sunderland Marina
0191 514 4721
Ch M, 80 H24
R Wear
Seaham
Hartlepool Marina
01429 865744
Ch M, 80
Hartlepool
R Tees
Whitby Marina
01947 600165
Ch 11 HW±2
Whitby
Robin Hood Bay
Scarborough
N
Bridlington
Hull Marina
01482 609960
Ch 80 HW±3
R Humber
South Ferriby Marina
01652 635620 Ch 80 HW±3
Meridian Quay Marina
01472 268424 Ch 74 HW±2
Boston Marina
01205 364420
HW±2
Wells
Wisbech Yacht Harbour
01945 588059
Ch 9 HW±3
Kings Lynn

WISBECH YACHT HARBOUR

Wisbech Yacht Harbour
Harbour Master, Harbour Office, The Boathouse,
Harbour Square, Wisbech, Cambridgeshire PE13 3BH
Tel: 01945 588059 Fax: 01945 580589
Email: afoster@fenland.gov.uk www.fenland.gov.uk

VHF	Ch 9
ACCESS	HW±3

Regarded as the capital of the English Fens, Wisbech is situated about 25 miles north east of Peterborough and is a market town of considerable character and historical significance. Rows of elegant houses line the banks of the River Nene, with the North and South Brink still deemed two of the finest Georgian streets in England.

Wisbech Yacht Harbour, linking Cambridgeshire with the sea, is proving increasingly popular as a haven for small craft, despite the busy commercial shipping. In recent years the facilities have been developed and improved upon and the HM is always on hand to help with passage planning both up or downstream.

FACILITIES AT A GLANCE

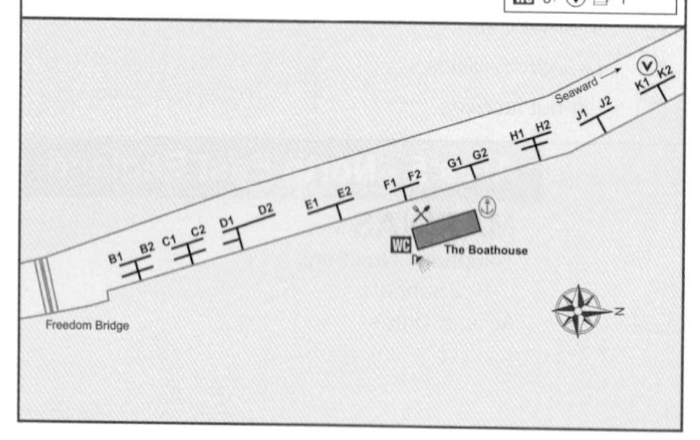
Seaward
K1 K2
J1 J2
H1 H2
G1 G2
F1 F2
E1 E2
D1 D2
C1 C2
B1 B2
The Boathouse
Freedom Bridge
N

BOSTON GATEWAY MARINA

Boston Gateway Marina
Witham Bank East, Boston, Lincs, PE21 9JU
Tel: 07480 525230
Email: enquiries@bostongatewaymarina.co.uk

VHF	
ACCESS	H±2

Located near Boston Grand Sluice Lock on the sunny side of the River Witham, the marina is ideally situated for easy access to The Wash and is suitable for both sea-going and river boats. It is a short walk to the centre of the historic town of Boston, Lincolnshire, but retains a tranquil feel. The marina offers visitor, short-term and longer-term moorings to suit each individual customer. Power and water are available for all boats.

The town centre offers the normal variety of facilities within easy walking distance. Local tourist attractions include the 14th century St Botolph's Church – 'The Stump' – the 1390s Boston Guildhall Musuem and a 450-year old market to name but a few.

FACILITIES AT A GLANCE

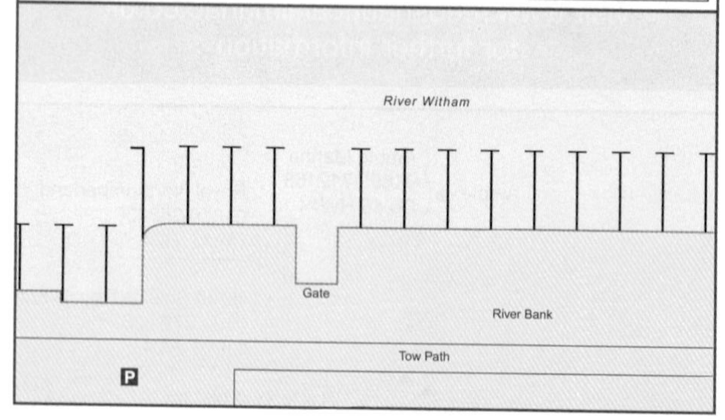
River Witham
Gate
River Bank
Tow Path
P

HUMBER CRUISING ASSOCIATION

Humber Cruising Association
Fish Docks, Grimsby, DN31 3SD
Tel: 01472 268424
www.hcagrimsby.co.uk
Email: berthmaster@hcagrimsby.co.uk

VHF	Ch 74
ACCESS	HW±2

Situated in the locked fish dock of Grimsby, at the mouth of the River Humber, Meridian Quay Marina is run by the Humber Cruising Association and comprises approximately 200 alongside berths plus 30 more for visitors. Accessed two hours either side of high water via lock gates, the lock should be contacted on VHF Ch 74 (call sign 'Fish Dock Island') as you make your final approach. The pontoon berths are equipped with water and electricity; there is a fully licensed clubhouse. Also available are internet access and laundry facilities.

FACILITIES AT A GLANCE

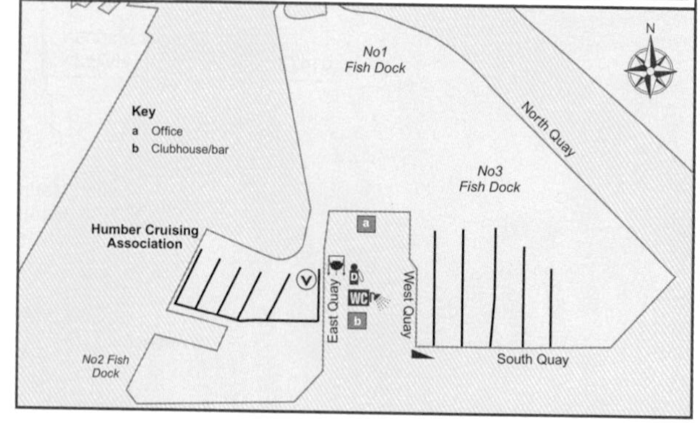
N
No1
Fish Dock
North Quay
Key
a Office
b Clubhouse/bar
No3
Fish Dock
Humber Cruising
Association
East Quay
West Quay
South Quay
No2 Fish
Dock

HULL WATERSIDE & MARINA

Hull Waterside & Marina
W 13, Kingston Street, Hull, HU1 2DQ
Tel: 01482 609960 Fax: 01482 224148
Email: hull@aquavista.com
www.aquavista.com

VHF	Ch 80
ACCESS	HW±3

Situated on the River Humber, Hull Marina is literally a stone's throw from the bustling city centre with its array of arts and entertainments. Besides the numerous historic bars and cafés surrounding the marina, there are plenty of traditional taverns to be sampled in the Old Town, while also found here is the Street Life Museum, vividly depicting the history of the city.

Yachtsmen enter the marina via a tidal lock, operating HW±3, and should try to give 15 minutes' notice of arrival via VHF Ch 80. Hull is perfectly positioned for exploring the Trent, Ouse and the Yorkshire coast as well as across the North Sea to Holland or Belgium.

FACILITIES AT A GLANCE

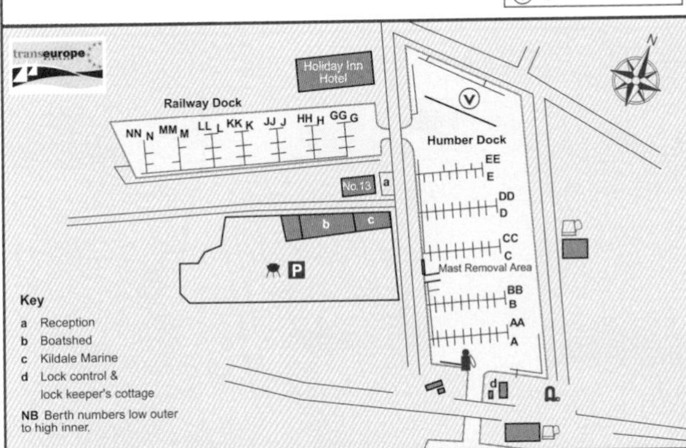
transeurope
Holiday Inn Hotel
Railway Dock
NN N MM M LL L KK K JJ J HH H GG G
Humber Dock
EE E
DD D
CC C
Mast Removal Area
BB B
AA A
No.13
a
b
c
P
Key
a Reception
b Boatshed
c Kildale Marine
d Lock control & lock keeper's cottage
NB Berth numbers low outer to high inner.

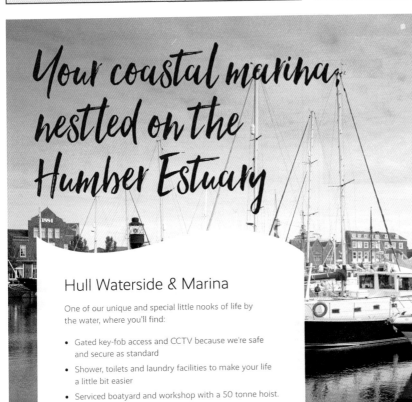

SOUTH FERRIBY MARINA

South Ferriby Marina
Red Lane, South Ferriby, Barton on Humber, Lincs, DN18 6JH
Tel: 01652 635620 (Lock 635219) Mobile: 07828 312071
Email: enquiries@southferribymarina.com

VHF	Ch 74
ACCESS	HW±3

Situated at the entrance to the non-tidal River Ancholme the existing marina has been established since 1966 and is well placed to provide easy access to the River Humber and North Sea. This is a family run business providing a range of services including a boatyard and chandlery. Access is by way of lock at HW±3.

The marina has excellent road and rail services within easy reach, while South Ferriby village has two pubs and a Post Office/Spar Shop just a short walk away.

The picturesque River Ancholme is navigable for about 17 miles; the maximum headroom under bridges is 4.42 metres (14ft 6ins).

FACILITIES AT A GLANCE

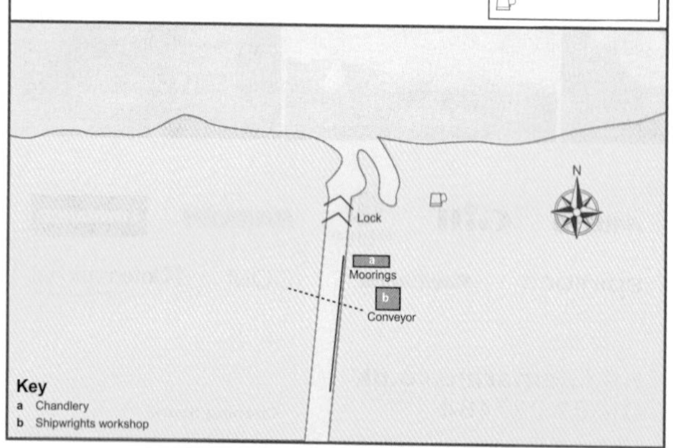
N
Lock
a
Moorings
b
Conveyor
Key
a Chandlery
b Shipwrights workshop

WHITBY MARINA

Whitby Marina
Whitby Harbour Office, Endeavour Wharf
Whitby, North Yorkshire YO21 1DN
Harbour Office: 01947 602354 Marina: 01947 600165
Email: port.services@scarborough.gov.uk

VHF	Ch 11
ACCESS	HW±2

The only natural harbour between the Tees and the Humber, Whitby lies some 20 miles north of Scarborough on the River Esk. The historic town is said to date back as far as the Roman times, although it is better known for its abbey, which was founded over 1,300 years ago by King Oswy of Northumberland. Another place of interest is the Captain Cook Memorial Museum, a tribute to Whitby's greatest seaman.

A swing bridge divides the harbour into upper and lower sections, with the marina being in the Upper Harbour. The bridge opens on request (VHF Ch 11) each half hour for two hours either side of high water.

FACILITIES AT A GLANCE

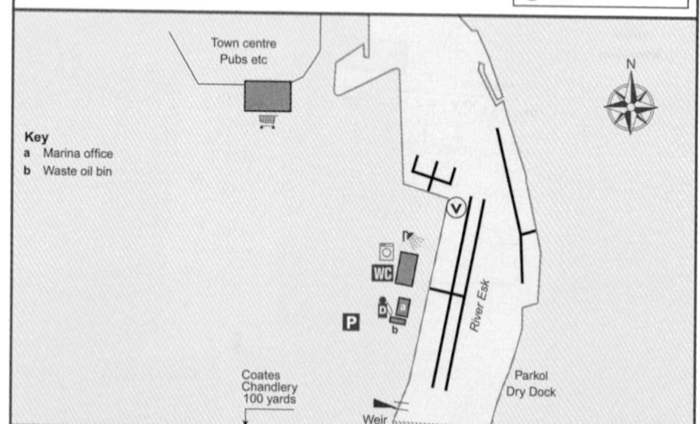
Town centre
Pubs etc
N
Key
a Marina office
b Waste oil bin
WC
P
River Esk
Coates
Chandlery
100 yards
Weir
Parkol
Dry Dock

HARTLEPOOL MARINA

Hartlepool Marina
Lock Office, Slake Terrace, Hartlepool, TS24 0RU
Tel: 01429 865744 www.hartlepool-marina.com
Email: enquiries@hartlepool-marina.com

VHF	Ch M, 80
ACCESS	

Hartlepool Marina is a modern boating facility on the NE coast now boasting an extensively refurbished North amenity block. Nestling on the Tees Valley the multi award winning marina promotes up to 500 pontoon berths alongside a variety of reputable services all surrounded by an exciting array of on water activities, a cosmopolitan mix of bistros, bars, restaurants, shopping, hotels and entertainment options.

Beautiful cruising waters and golden sands to the North and South of the marina approach which is channel dredged to CD and accessible via a lock: vessels wishing to enter should contact the Marina Lock Office on VHF Ch M/80 before arrival.

FACILITIES AT A GLANCE

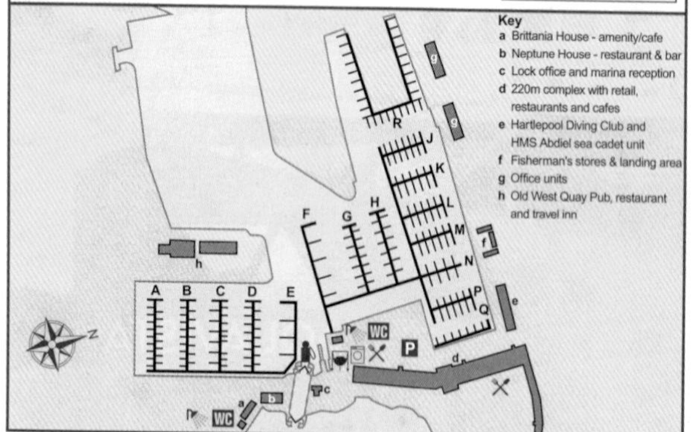
Key
a Brittania House - amenity/cafe
b Neptune House - restaurant & bar
c Lock office and marina reception
d 220m complex with retail, restaurants and cafes
e Hartlepool Diving Club and HMS Abdiel sea cadet unit
f Fisherman's stores & landing area
g Office units
h Old West Quay Pub, restaurant and travel inn
A B C D E F G H J K L M N P Q R
N
WC
P

SUNDERLAND MARINA

The Marine Activities Centre
Sunderland Marina, Sunderland, SR6 0PW
Tel: 0191 514 4721
Email: info@sunmac.org.uk

VHF	Ch M
ACCESS	H24

Sunderland Marina sits on the the River Wear and is easily accessible through the outer breakwater at all states of tide. A short walk away from the city centre and beautiful beaches, facilities on site include the Snowgoose café and the Marina Vista Italian restaurant. Other pubs, restaurants, hotels and cafes are located nearby on the waterfront.

Sunderland Yacht Club is also located nearby and welcomes visiting yachtsman to its clubhouse.

FACILITIES AT A GLANCE

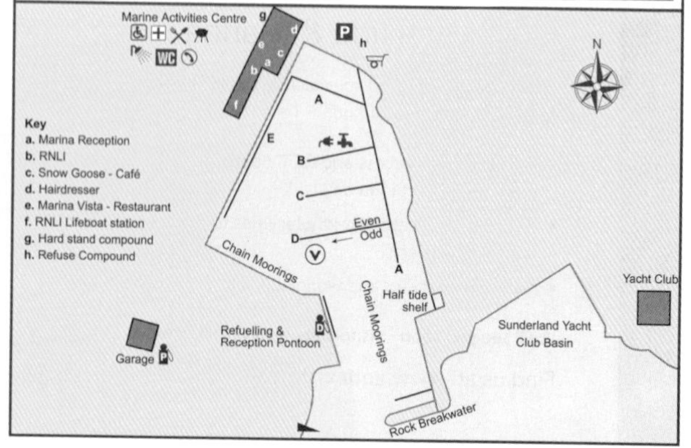
Marine Activities Centre
WC
P
N
Key
a. Marina Reception
b. RNLI
c. Snow Goose - Café
d. Hairdresser
e. Marina Vista - Restaurant
f. RNLI Lifeboat station
g. Hard stand compound
h. Refuse Compound
A
B
C
D
E
Even
Odd
Chain Moorings
Half tide shelf
Garage
Refuelling & Reception Pontoon
Sunderland Yacht Club Basin
Yacht Club
Rock Breakwater

ROYAL QUAYS MARINA

Royal Quays Marina
Coble Dene Road, North Shields, NE29 6DU
Tel: 0191 272 8282 Fax: 0191 272 8288
Email: royalquays@boatfolk.co.uk
www.boatfolk.co.uk/royalquays

VHF Ch 80
ACCESS H24

Royal Quays Marina enjoys close proximity to the entrance to the River Tyne, allowing easy access to and from the open sea as well as being ideally placed for cruising further up the Tyne. Just over an hour's motoring upstream brings you to the heart of the city of Newcastle, where you can tie up on a security controlled visitors' pontoon right outside the Pitcher and Piano Bar.

With a reputation for a high standard of service, the marina accommodates 300 pontoon berths, all of which are fully serviced. It is accessed via double sector lock gates which operate at all states of the tide and 24 hours a day.

FACILITIES AT A GLANCE

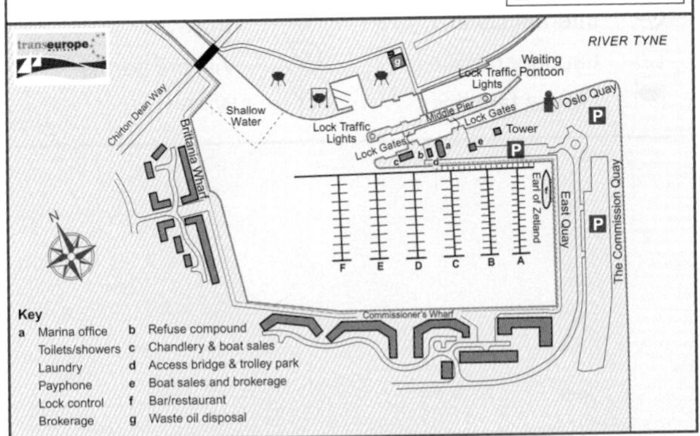
RIVER TYNE
Waiting Pontoon
Lock Traffic Lights
Shallow Water
Lock Traffic Lights
Lock Gates
Middle Pier
Lock Gates
Oslo Quay
Tower
Chirton Dean Way
Britannia Wharf
Earl of Zetland
East Quay
The Commission Quay
F E D C B A
Commissioner's Wharf
N
Key
a Marina office
Toilets/showers
Laundry
Payphone
Lock control
Brokerage
b Refuse compound
c Chandlery & boat sales
d Access bridge & trolley park
e Boat sales and brokerage
f Bar/restaurant
g Waste oil disposal

ST PETERS MARINA

St Peters Marina, St Peters Basin
Newcastle upon Tyne, NE6 1HX
Tel: 0191 2654472 Fax: 0191 2762618
Email: info@stpetersmarina.co.uk
www.stpetersmarina.co.uk

VHF Ch 80
ACCESS HW±3

Nestling on the north bank of the River Tyne, some eight miles upstream of the river entrance, St Peters Marina is a fully serviced, 150-berth marina with the capacity to accommodate large vessels of up to 37m LOA. Situated on site is the Bascule Bar and Bistro, while a few minutes away is the centre of Newcastle. This city, along with its surrounding area, offers an array of interesting sites, among which are Hadrian's Wall, the award winning Gateshead Millennium Bridge and the Baltic Art Centre.

FACILITIES AT A GLANCE

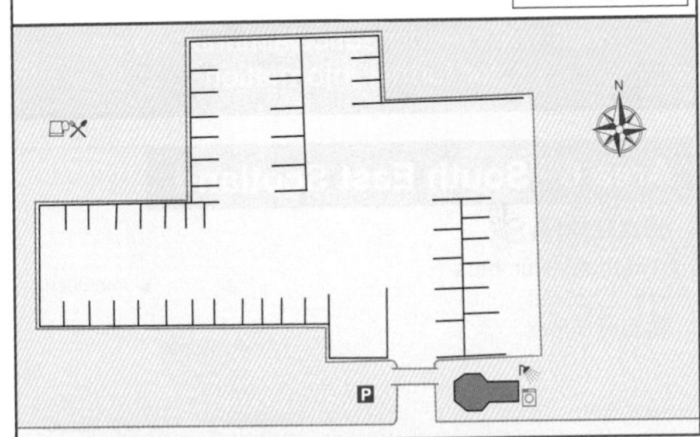
N

ROYAL NORTHUMBERLAND YACHT CLUB

Royal Northumberland Yacht Club
South Harbour, Blyth, Northumberland, NE24 3PB
Tel: 01670 353636

VHF Ch 12
ACCESS H24

The Royal Northumberland Yacht Club is based at Blyth, a well-sheltered port that is accessible at all states of the tide and in all weathers except for when there is a combination of low water and strong south-easterly winds. The yacht club is a private club with some 75 pontoon berths and a further 20 fore and aft moorings.

Visitors usually berth on the north side of the most northerly pontoon and are welcome to use the clubship, HY *Tyne* – a wooden lightship built in 1880 which incorporates a bar, showers and toilet facilities. The club also controls its own boatyard, providing under cover and outside storage space plus a 20 ton boat hoist.

FACILITIES AT A GLANCE

South Harbour
West Quay
Mast
Middle Jetty
Ro Ro
H.Y. Tyne
Outer West Pier
N

AMBLE MARINA

Amble Marina Ltd
Amble, Northumberland, NE65 0YP
Tel: 01665 712168
Email: marina@amble.co.uk www.amble.co.uk

VHF Ch 80
ACCESS HW±4

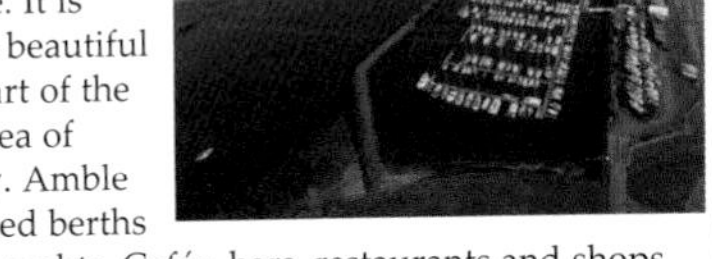

Amble Marina is a small family run business offering peace, security and a countryside setting at the heart of the small town of Amble. It is located on the banks of the beautiful River Coquet and at the start of the Northumberland coast's area of outstanding natural beauty. Amble Marina has 250 fully serviced berths for residential and visiting yachts. Cafés, bars, restaurants and shops are all within a short walk.

From your berth watch the sun rise at the harbour entrance and set behind Warkworth Castle or walk on wide, empty beaches. There is so much to do or if you prefer simply enjoy the peace, tranquillity and friendliness at Amble Marina.

FACILITIES AT A GLANCE

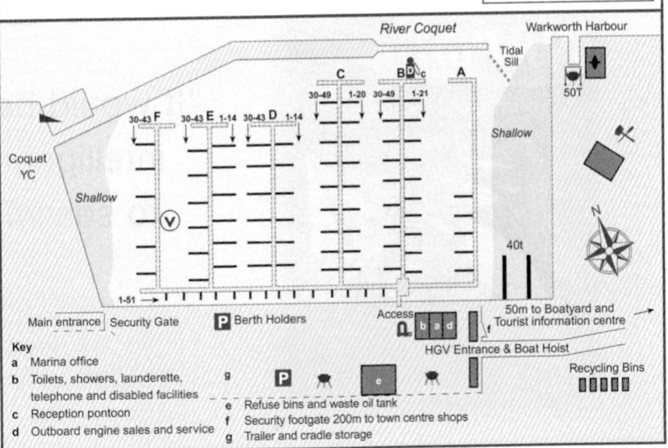
River Coquet
Warkworth Harbour
Tidal Sill
50T
Shallow
Coquet YC
Shallow
40t
Main entrance
Security Gate
Berth Holders
Access
50m to Boatyard and Tourist information centre
HGV Entrance & Boat Hoist
Recycling Bins
N
Key
a Marina office
b Toilets, showers, launderette, telephone and disabled facilities
c Reception pontoon
d Outboard engine sales and service
e Refuse bins and waste oil tank
f Security footgate 200m to town centre shops
g Trailer and cradle storage

SOUTH EAST SCOTLAND – Eyemouth to Rattray Head

Key to Marina Plans symbols

Bottled gas	Parking
Chandler	Pub/Restaurant
Disabled facilities	Pump out
Electrical supply	Rigging service
Electrical repairs	Sail repairs
Engine repairs	Shipwright
First Aid	Shop/Supermarket
Fresh Water	Showers
Fuel - Diesel	Slipway
Fuel - Petrol	Toilets
Hardstanding/boatyard	Telephone
Internet Café	Trolleys
Laundry facilities	Visitors berths
Lift-out facilities	Wi-Fi

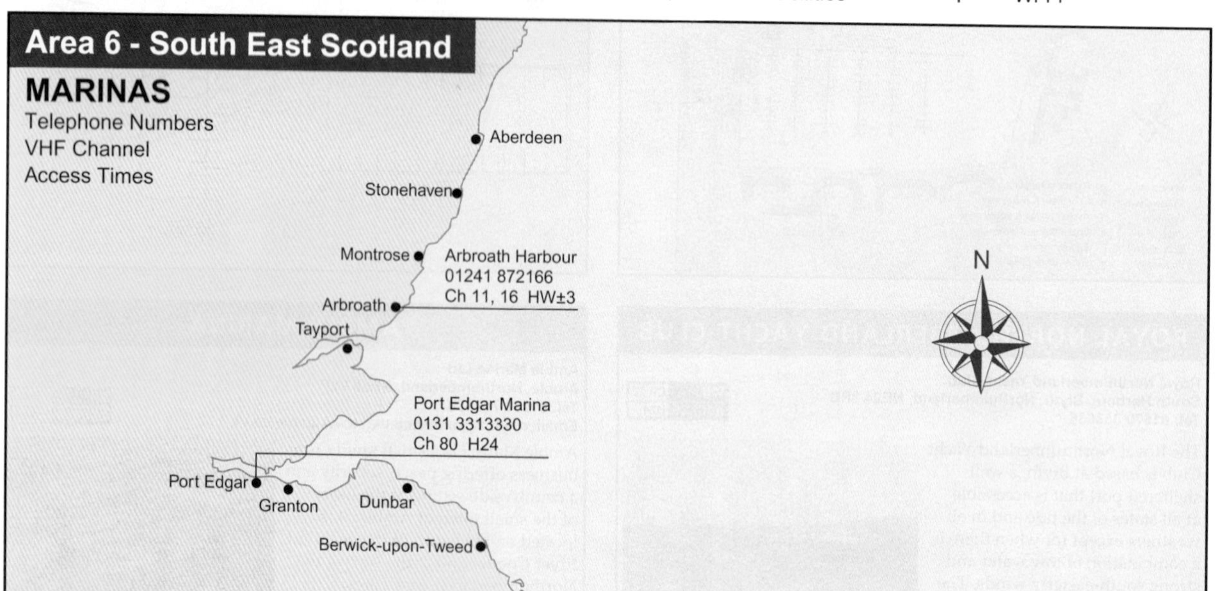
Area 6 - South East Scotland
MARINAS
Telephone Numbers
VHF Channel
Access Times
Aberdeen
Stonehaven
Montrose
Arbroath Harbour
01241 872166
Ch 11, 16 HW±3
Arbroath
Tayport
Port Edgar Marina
0131 3313330
Ch 80 H24
Port Edgar
Granton
Dunbar
Berwick-upon-Tweed
N

PORT EDGAR MARINA

Port Edgar Marina
Shore Road, South Queensferry
West Lothian, EH30 9SQ
Tel: 0131 331 3330 Fax: 0131 331 4878
Email: info@portedgar.co.uk

VHF	Ch 80
ACCESS	H24

Nestled between the iconic Forth Bridges, Edinburgh's 300 berth marina is the ideal base for exploring the Capital and the Forth coastline.

A short walk away is the historic High Street of Queensferry with a great selection of bars and restaurants. Situated 15 minutes away from Edinburgh Airport with easy road access, the secure site provides full boatyard facilities including a 25T slipway hoist, chandlery and café.

FACILITIES AT A GLANCE

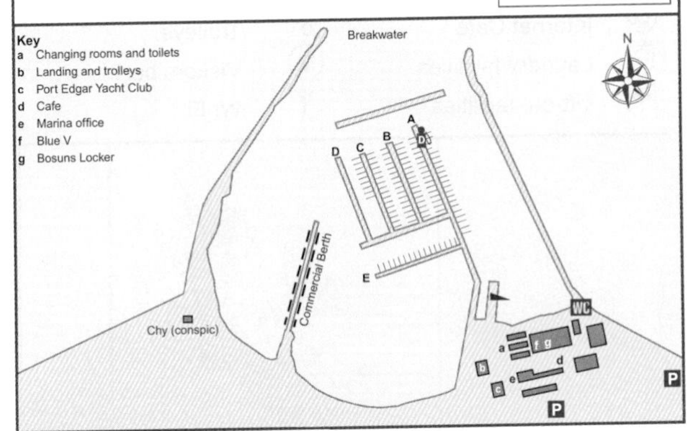
Key
a Changing rooms and toilets
b Landing and trolleys
c Port Edgar Yacht Club
d Cafe
e Marina office
f Blue V
g Bosuns Locker
Breakwater
Commercial Berth
Chy (conspic)
A
B
C
D
E
WC
P

ARBROATH HARBOUR

Arbroath Harbour
Harbour Office, Arbroath, DD11 1PD
Tel: 01241 872166 Fax: 01241 878472
Email: harbourmaster@angus.gov.uk

VHF	Ch 11
ACCESS	HW±3

GEMINI

Arbroath harbour has 59 floating pontoon berths with security entrance which are serviced with electricity and fresh water to accommodate all types of leisure craft. Half height dock gates with walkway are located between the inner and outer harbours, which open and close at half tide, maintaining a minimum of 2.5m of water in the inner harbour.

The town of Arbroath offers a variety of social and sporting amenities to visiting crews and a number of quality pubs, restaurants, the famous twelfth century Abbey and Signal Tower Museum are located close to the harbour. Railway and bus stations are only 1km from the harbour with direct north and south connections.

FACILITIES AT A GLANCE

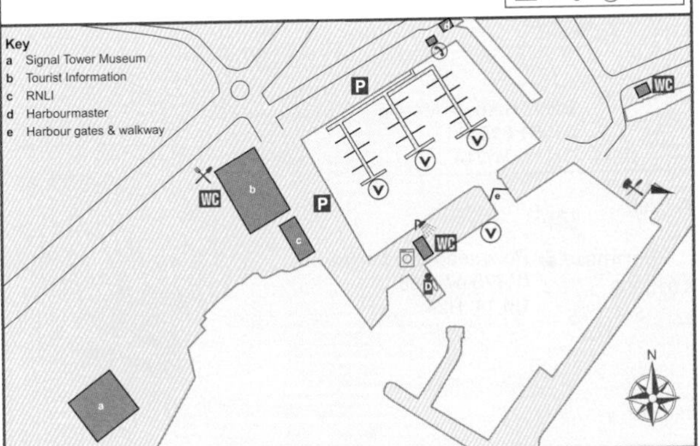
Key
a Signal Tower Museum
b Tourist Information
c RNLI
d Harbourmaster
e Harbour gates & walkway
P
WC
N

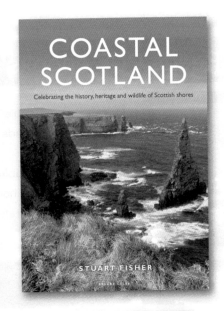

6

Key to Marina Plans symbols

- Bottled gas
- Chandler
- Disabled facilities
- Electrical supply
- Electrical repairs
- Engine repairs
- First Aid
- Fresh Water
- Fuel - Diesel
- Fuel - Petrol
- Hardstanding/boatyard
- Internet Café
- Laundry facilities
- Lift-out facilities
- Parking
- Pub/Restaurant
- Pump out
- Rigging service
- Sail repairs
- Shipwright
- Shop/Supermarket
- Showers
- Slipway
- Toilets
- Telephone
- Trolleys
- Visitors berths
- Wi-Fi

Area 7 - North East Scotland

MARINAS

Telephone Numbers
VHF Channel
Access Times

Shetland Islands

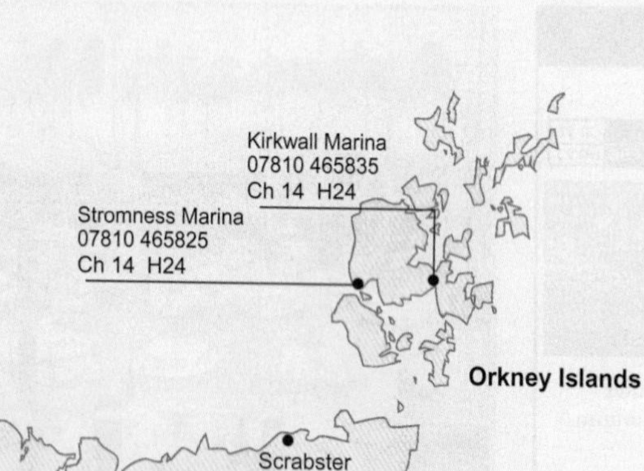
Kirkwall Marina
07810 465835
Ch 14 H24
Stromness Marina
07810 465825
Ch 14 H24
Orkney Islands
Scrabster
Wick
Wick Marina
01955 602030
Ch 14 H24
Helmsdale
Ullapool
Whitehills Marina
01261 861291
Ch 14 H4
Banff Harbour Marina
01261 815544
Ch 12 HW±4
Inverness Marina
07526 446348
Ch 12
Inverness
Buckie
Banff
Macduff
Findhorn
Burghead
Caley Marina
01463 236539
Ch 74 H24
Seaport Marina
01463 725500
Ch 74 HW±4
Lossiemouth 01343 813066
Ch 12 HW±4
Nairn Marina
01667 456008
Ch 10 HW±2
Hopeman
Peterhead
Peterhead Bay Marina
01779 477868
Ch 14 H24
Mallaig
Aberdeen

N

PETERHEAD BAY MARINA

Peterhead Port Authority
Harbour Office, West Pier, Peterhead, AB42 1DW
Tel: 01779 477868/483600
Email: marina@peterheadport.co.uk
www.peterheadport.co.uk

VHF Ch 14
ACCESS H24

Based in the south west corner of Peterhead Bay Harbour, the marina provides one of the finest marine leisure facilities in the east of Scotland. In addition to the services on site, there are plenty of nautical businesses in the vicinity, ranging from ship chandlers and electrical servicing to boat repairs and surveying.

Due to its easterly location, Peterhead affords an ideal stopover for those yachts heading to or from Scandinavia as well as for vessels making for the Caledonian Canal.

FACILITIES AT A GLANCE

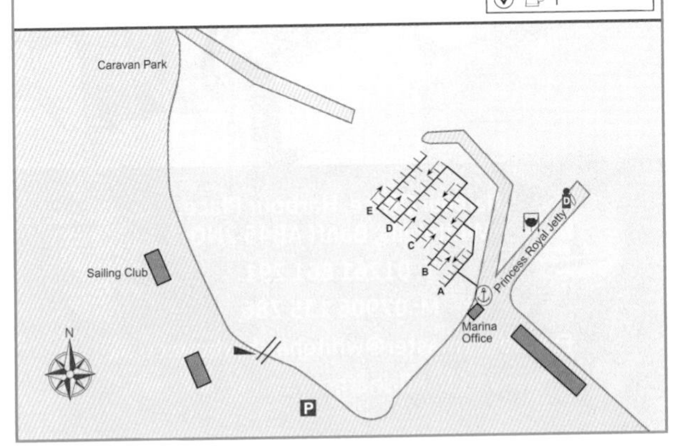
Caravan Park
Sailing Club
Princess Royal Jetty
Marina Office
A
B
C
D
E
N
P

7

BANFF HARBOUR MARINA

Banff Harbour Marina
Harbour Office, Quayside, Banff, Aberdeenshire, AB45 1HQ
Tel: 01261 832236 Mobile: 07920 270360
Email: duncan.mackie@aberdeenshire.gov.uk

VHF Ch 12
ACCESS HW±4

A former fishing and cargo port now used as a recreational harbour. Banff offers excellent facilities to both regular and visiting users. The marina now provides 92 berths, of which 76 are serviced pontoon berths and 16 unserviced, traditional moorings, in one of the safest harbours on the NE coast of Scotland.

The outer basin offers adequate berthing for visitors and a tidal area for regulars.

The harbour is tidal with a sandy bottom. Movement during low water neaps is no problem for the shallow drafted boat.

FACILITIES AT A GLANCE

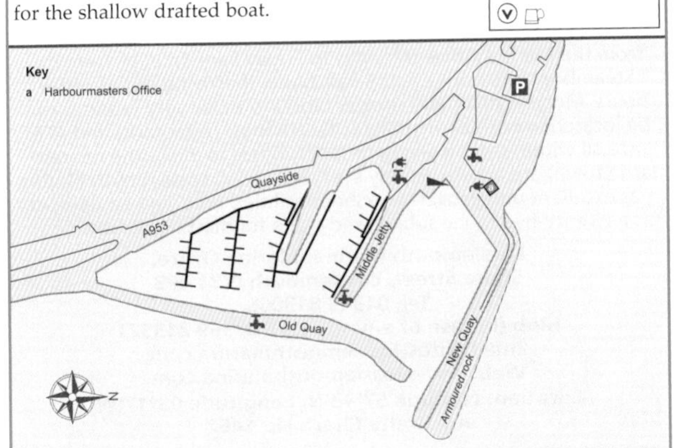
Key
a Harbourmasters Office
Quayside
A953
Middle Jetty
Old Quay
New Quay
Armoured rock
P
N

NAIRN MARINA

Nairn Marina
Nairn Harbour, Nairnshire, Scotland
Tel: 01667 456008 Fax: 01667 452877
Email: nairn.harbourmaster@virgin.net

VHF Ch 10
ACCESS HW±2

Nairn is a small town on the coast of the Moray Firth. Formerly renowned both as a fishing port and as a holiday resort dating back to Victorian times, it boasts miles of award-winning, sandy beaches, famous castles such as Cawdor, Brodie and Castle Stuart, and two championship golf courses. Other recreational activities include horse riding or walking through spectacular countryside.

The marina lies at the mouth of the River Nairn, entry to which should be avoided in strong N to NE winds. The approach is made from the NW at or around high water as the entrance is badly silted and dries out.

FACILITIES AT A GLANCE

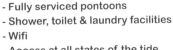

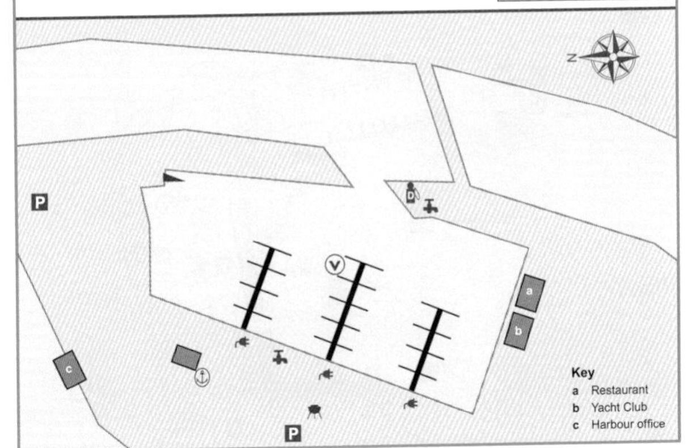
Key
a Restaurant
b Yacht Club
c Harbour office
P
N

WHITEHILLS MARINA

Whitehills Harbour Commissioners
Whitehills, Banffshire AB45 2NQ
Tel: 01261 861291
www.whitehillsharbour.co.uk
Email: harbourmaster@whitehillsharbour.co.uk

VHF	Ch 14
ACCESS	H24

Built in 1900, Whitehills is a Trust Harbour fully maintained and run by nine commissioners elected from the village. It was a thriving fishing port up until 1999, but due to changes in the fishing industry, was converted into a marina during 2000.

Photo by Colin Heggie

Three miles west of Banff Harbour the marina benefits from good tidal access – although there is just 1.5m at springs – comprising 38 serviced berths, with electricity, as well as eight non-serviced berths.

Whitehills village has a wide range of facilities including a convenience store, a cafe/fish & chip shop, two pubs, a fresh fish shop as well as two good restaurants. It is also a great base for families, with an excellent playpark at Blackpots, just a short walk from the harbour.

FACILITIES AT A GLANCE

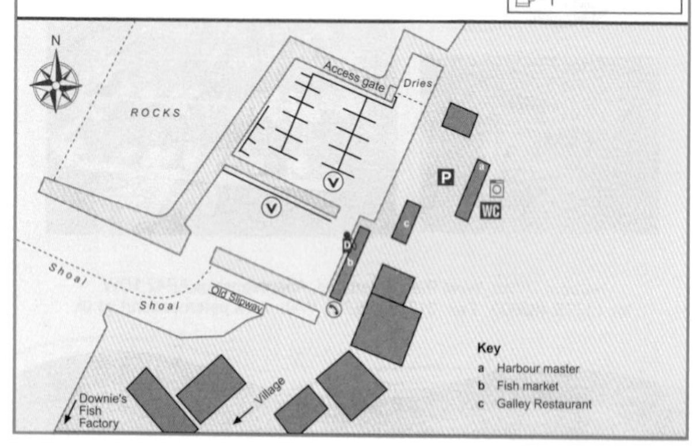
N
ROCKS
Access gate
Dries
Shoal
Shoal
Old Slipway
Village
Downie's Fish Factory
Key
a Harbour master
b Fish market
c Galley Restaurant

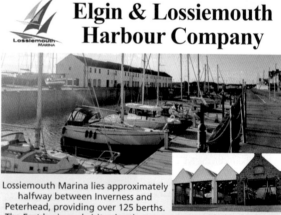

LOSSIEMOUTH MARINA

Marina Office
Lossiemouth, Moray, IV31 6PB
Tel: 01343 813066
Email: info@lossiemouthmarina.com

VHF	Ch 12
ACCESS	HW±4

Approximately halfway between Inverness and Peterhead, the marina provides over 115 berths. The East Basin and visitor berth area were upgraded in 2018 and the dedicated visitor finger pontoons lie just inside the East Basin, providing free water and electricity. Visitor packs can be collected from the marina office (Mon–Fri 9am-4pm) or from the Steamboat Inn. Modern toilet and shower blocks with laundry facilities are located in both basins. Diesel, local shops, restaurants and ATM are all within short walking distance. The marina has excellent undercover workshop facilities, dredging equipment, a 25 tonne sublift, and crane for masting/demasting.

FACILITIES AT A GLANCE

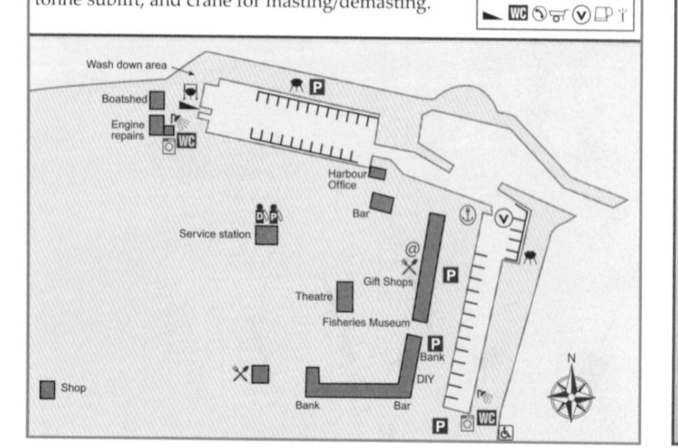
Wash down area
Boatshed
Engine repairs
Harbour Office
Bar
Service station
Gift Shops
Theatre
Fisheries Museum
Bank
DIY
Shop
Bank
Bar
N

INVERNESS MARINA

Inverness Marina
Longman Drive, Inverness, IV1 1SU
Tel: 01463 220501
Email: info@invernessmarina.com
www.invernessmarina.com

VHF	Ch 12
ACCESS	H24

The marina is situated in the Inverness firth just one mile from the city centre and half a mile from the entrance to the Caledonian Canal. It has a minimum depth of 3m, 24hr access and 150 fully serviced berths. On site are a chandlery and services including rigging, engineering, electronics and boat repair.

Inverness has excellent transport networks to the rest of the UK and Europe and, as the gateway to the Highlands is a great location as a base for a touring golf courses, historic sites and the Whisky Trail. The marina is a perfect base for cruising Orkney, Shetland and Scandinavia.

FACILITIES AT A GLANCE

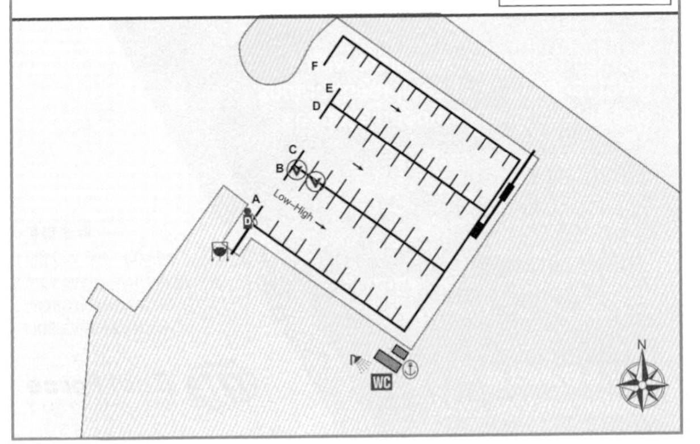
F
E
D
C
B
A
Low–High
WC
N

SEAPORT MARINA

Seaport Marina
Muirtown Wharf, Inverness, IV3 5LE
Tel: 01463 725500
Email: caledonian@scottishcanals.co.uk
www.scottishcanals.co.uk

VHF	Ch 74
ACCESS	HW±4

Photo courtesy of D Edes

Seaport Marina is based at Muirtown Basin at the eastern entrance of the Caledonian Canal; a 60 mile coast-to-coast channel slicing through the majestic Great Glen. Only a 15 minute walk from the centre of Inverness, the Marina is an ideal base for visiting the Highlands.

There are shops and amenities nearby, as well as chandlers, boat repair services and a slipway. The marina also offers a variety of winter mooring packages and details of transit and short term licences, including the use of the Caledonian Canal can be found on the above website.

FACILITIES AT A GLANCE

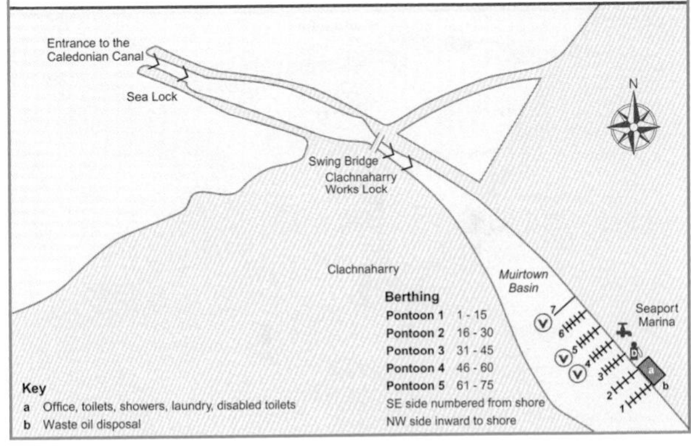
Entrance to the Caledonian Canal
Sea Lock
Swing Bridge
Clachnaharry Works Lock
Clachnaharry
Muirtown Basin
Seaport Marina
N
Berthing
Pontoon 1 1 - 15
Pontoon 2 16 - 30
Pontoon 3 31 - 45
Pontoon 4 46 - 60
Pontoon 5 61 - 75
SE side numbered from shore
NW side inward to shore
Key
a Office, toilets, showers, laundry, disabled toilets
b Waste oil disposal

7

CALEY MARINA

Caley Marina
Canal Road, Inverness, IV3 8NF
Tel: 01463 236539 Fax: 01463 238323
Email: info@caleymarina.com
www.caleymarina.com

VHF	Ch 74
ACCESS	H24

Caley Marina is a family run business based near Inverness. With the four flight Muirtown locks and the Kessock Bridge providing a dramatic backdrop, the marina runs alongside the Caledonian Canal which, opened in 1822, is regarded as one of the most spectacular waterways in Europe. Built as a short cut between the North Sea and the Atlantic Ocean, thus avoiding the potentially dangerous Pentland Firth on the north coast of Scotland, the canal is around 60 miles long and takes about three days to cruise from east to west. With the prevailing winds behind you, it takes slightly less time to cruise in the other direction.

FACILITIES AT A GLANCE

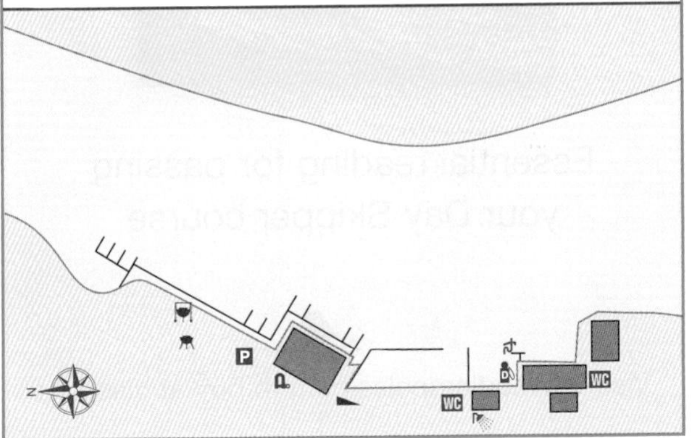
N
WC
WC
WC

WICK MARINA

Wick Marina
Harbour Office, Wick, Caithness, KW1 5HA
Tel: 01955 602030
Email:office@wickharbour.co.uk

VHF	Ch 14, 16
ACCESS	H24

This is the most northerly marina on the British mainland and the last stop before the Orkney and Shetland Islands. Situated an easy five minutes walk from the town centre Wick Marina accommodates 80 fully serviced berths with all the support facilities expected in a modern marina.

This part of Scotland with its rugged coastline and rich history is easily accessible by air and a great starting point for cruising in the northern isles, Moray Firth, Caledonian Canal and Scandinavia, a comfortable 280-mile sail.

FACILITIES AT A GLANCE

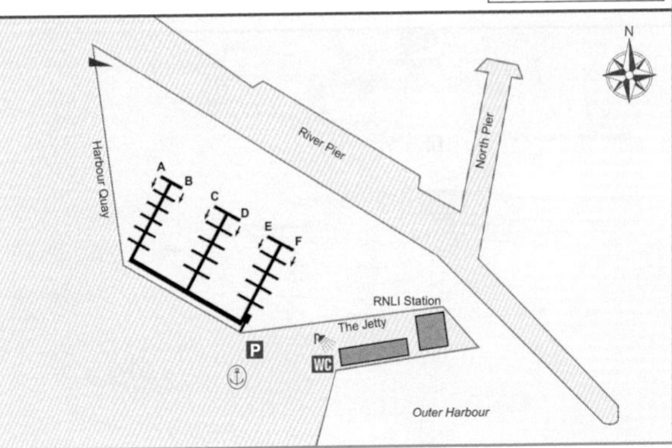
Harbour Quay
River Pier
North Pier
N
A
B
C
D
E
F
RNLI Station
The Jetty
WC
Outer Harbour

KIRKWALL MARINA

Kirkwall Marina
Harbour Street, Kirkwall, Orkney, KW15
Tel: 07810 465835 Fax: 01856 871313
Email: info@orkneymarinas.co.uk www.orkneymarinas.co.uk

VHF	Ch 14
ACCESS	H24

The Orkney Isles, comprising 70 islands in total, provides some of the finest cruising grounds in Northern Europe. The Main Island, incorporating the ancient port of Kirkwall, is the largest, although 16 others have lively communities and are rich in archaeological sites as well as spectacular scenery and wildlife.

Kirkwall Marina, an all year facility, is located within the harbour and just yards from the visitor attractions of this ancient port. Local shops, hotels and restaurants are all within walking distance.

FACILITIES AT A GLANCE

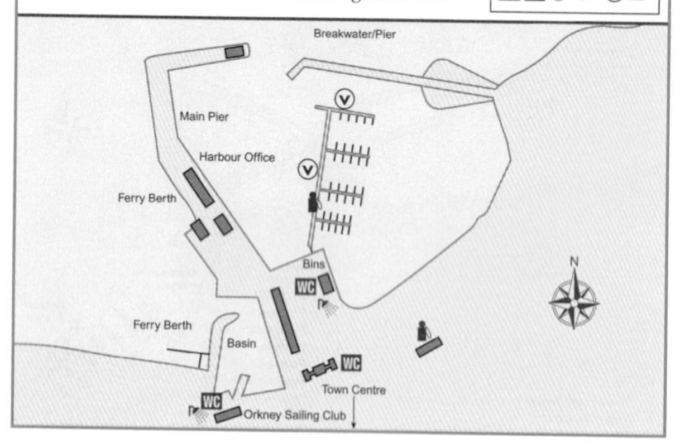
Breakwater/Pier
Main Pier
Harbour Office
Ferry Berth
Bins
Ferry Berth
Basin
Town Centre
Orkney Sailing Club
N

Reputation Built on Trust
Scotland's largest chandlers with a great range of boat equipment brands and servicing, at competitive prices.
Boat Maintenance and Paints
Moorings Services
Outboard Engine Servicing
Paddlesports Gear
Sailing and Commercial Clothing
Safety and Marine Hardware
+44 (0) 1463 229400
sales@gaelforcegroup.com
Anderson Street,
Inverness, IV3 8DH
Click & Collect at
gaelforcemarine.co.uk
Gael Force
MARINE EQUIPMENT

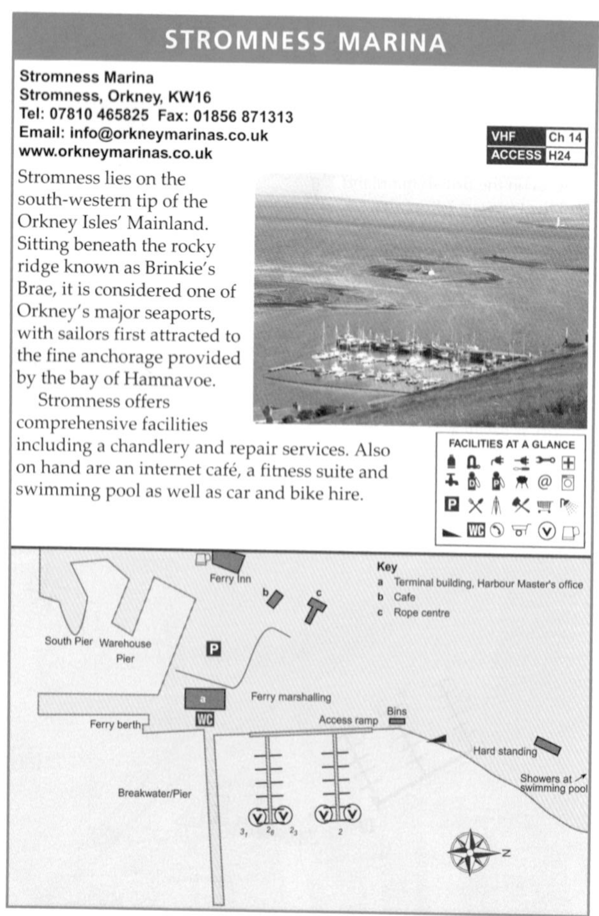

STROMNESS MARINA

Stromness Marina
Stromness, Orkney, KW16
Tel: 07810 465825 Fax: 01856 871313
Email: info@orkneymarinas.co.uk
www.orkneymarinas.co.uk

VHF	Ch 14
ACCESS	H24

Stromness lies on the south-western tip of the Orkney Isles' Mainland. Sitting beneath the rocky ridge known as Brinkie's Brae, it is considered one of Orkney's major seaports, with sailors first attracted to the fine anchorage provided by the bay of Hamnavoe.

Stromness offers comprehensive facilities including a chandlery and repair services. Also on hand are an internet café, a fitness suite and swimming pool as well as car and bike hire.

FACILITIES AT A GLANCE
Ferry Inn
Key
a Terminal building, Harbour Master's office
b Cafe
c Rope centre
South Pier
Warehouse Pier
Ferry marshalling
Ferry berth
Access ramp
Bins
Hard standing
Showers at swimming pool
Breakwater/Pier
N

DAY SKIPPER
EXERCISES FOR SAIL & POWER
ALISON NOICE
& ROGER SEYMOUR
2nd edition
ADLARD COLES
Essential reading for passing your Day Skipper course
Visit www.adlardcoles.com to buy at discount

NORTH WEST SCOTLAND – Cape Wrath to Crinan Canal

Key to Marina Plans symbols

Bottled gas	Parking
Chandler	Pub/Restaurant
Disabled facilities	Pump out
Electrical supply	Rigging service
Electrical repairs	Sail repairs
Engine repairs	Shipwright
First Aid	Shop/Supermarket
Fresh Water	Showers
Fuel - Diesel	Slipway
Fuel - Petrol	Toilets
Hardstanding/boatyard	Telephone
Internet Café	Trolleys
Laundry facilities	Visitors berths
Lift-out facilities	Wi-Fi

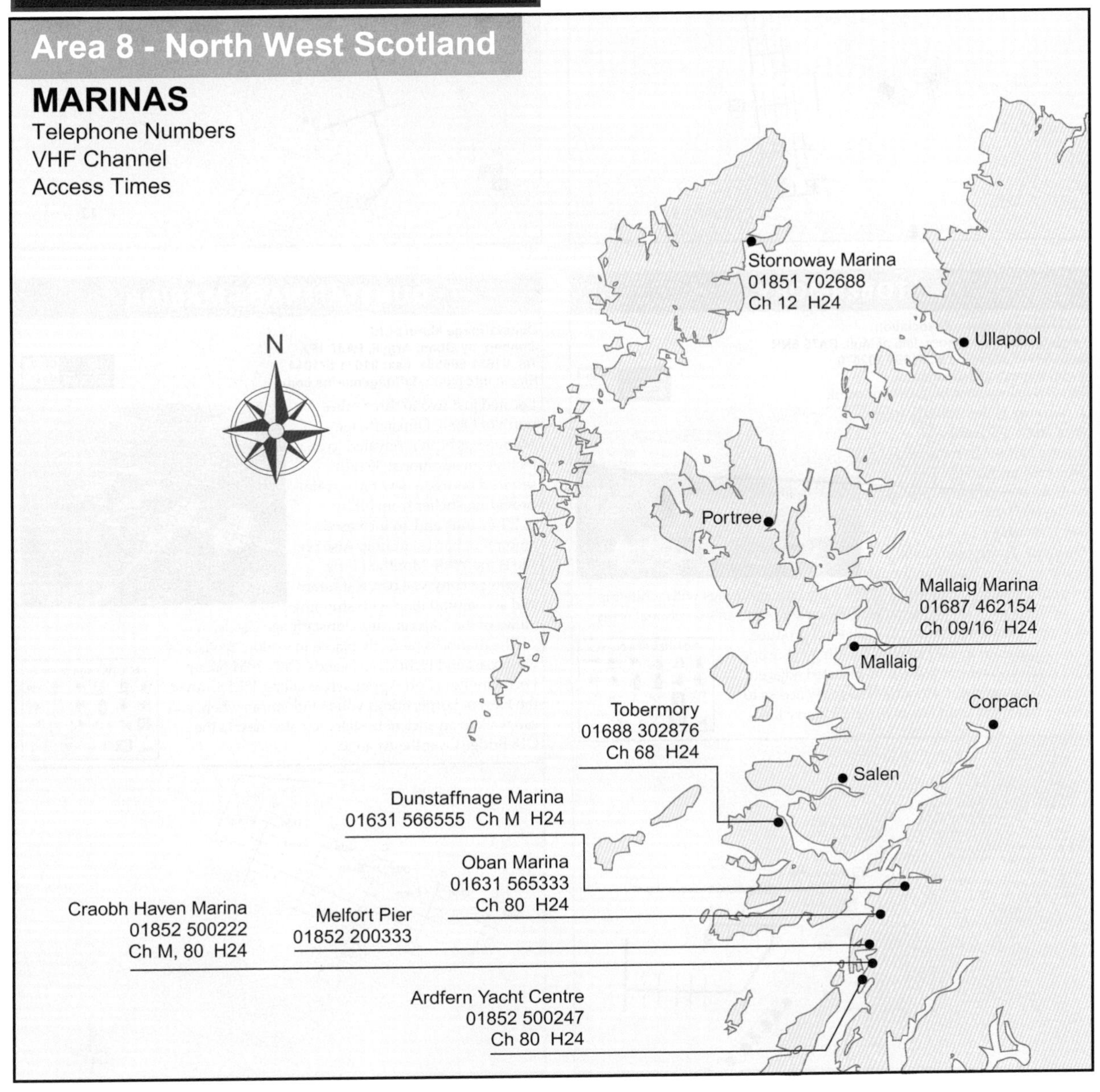
Area 8 - North West Scotland
MARINAS
Telephone Numbers
VHF Channel
Access Times
N
Stornoway Marina
01851 702688
Ch 12 H24
Ullapool
Portree
Mallaig Marina
01687 462154
Ch 09/16 H24
Mallaig
Corpach
Tobermory
01688 302876
Ch 68 H24
Salen
Dunstaffnage Marina
01631 566555 Ch M H24
Oban Marina
01631 565333
Ch 80 H24
Craobh Haven Marina
01852 500222
Ch M, 80 H24
Melfort Pier
01852 200333
Ardfern Yacht Centre
01852 500247
Ch 80 H24

STORNOWAY MARINA

Stornoway Port Authority
Amity House, Esplanade Quay,
Stornoway, Isle of Lewis, HS1 2XS
Tel: 01851 702688 Fax: 01851 705714
Email: sypa@stornowayport.com

VHF Ch 12
ACCESS HW24

Stornoway Marina is sheltered and has easy access at all states of the tide and weather conditions. Vessels up to 24 metres in length and 3 metres draft can be accommodated.

The 80-berth marina provides a safe haven for island hoppers and days sailors. The marina is particularly popular as it is located right in the heart of the bustling town centre. Fresh water, electricity, wi-fi, toilet, shower and laundry facilities are available quayside for all visitors.

FACILITIES AT A GLANCE

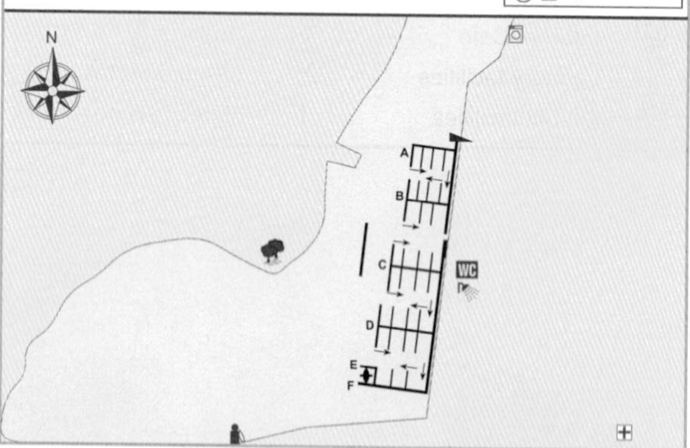
N
A
B
C
D
E
F
WC

MALLAIG MARINA

Mallaig Marina
East Bay, Mallaig, Inverness-shire, PH41 4QS
Tel: 07824 331031 Fax: 01687 462172
Email: info@mallaigharbourauthority.com

VHF Ch 09, 16
ACCESS H24

Mallaig Marina is truly the gateway to the Western Isles. The 50-berth marina, part funded by EU Sail West project, provides the ideal location for experiencing and exploring the magnificent sailing opportunities available on the west coast of Scotland. Onshore facilities including a small laundry are available in the Marina Centre.

The marina is a short walk from the village centre, which offers a range of options for the discerning diner, shopper or tourist. Fishing boats still operate from the busy harbour, and there is regular activity with various ferries to and fro Knoydart, the Small Isles and the Western Isles. The village also welcomes the Jacobite Steam Train twice daily in the high season.

FACILITIES AT A GLANCE

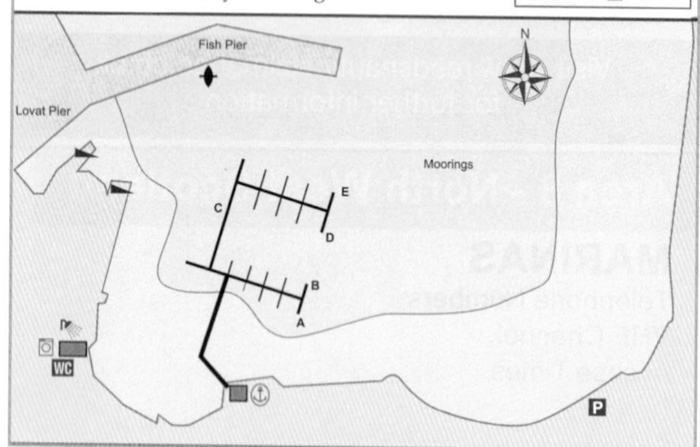
Fish Pier
Lovat Pier
N
Moorings
C
E
D
B
A
WC
P

TOBERMORY

Tobermory Harbour Association
Taigh Solais, Tobermory, Isle of Mull, PA75 6NR
Mob: 07917 832497 Tel: 01688 302876
www.tobermoryharbour.co.uk
Email: admin@tobermoryharbour.co.uk

VHF Ch 68
ACCESS H24

Tobermory is the iconic Scottish west coast destination, a natural historic harbour and protected anchorage. The harbour pontoons are located on the west shore of Tobermory Bay with direct access to the town. To supplement 50 pontoon berths there are also 30 swinging moorings for hire – look for the blue moorings with white top.

Within easy walking distance, Tobermory offers an exceptional array of shops, bars and restaurants with local produce. Mull Aquarium in the Harbour Building is Europe's first catch and release aquarium. Situated adjacent to the main car par, the pontoon has easy access to public transport to and from mainland ferry links.

FACILITIES AT A GLANCE

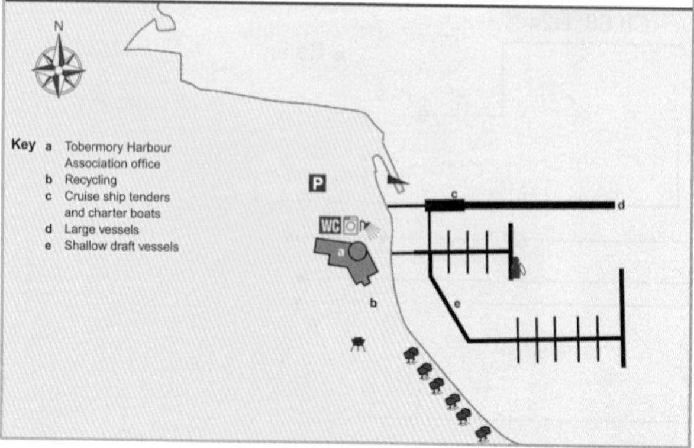
N
Key a Tobermory Harbour Association office
b Recycling
c Cruise ship tenders and charter boats
d Large vessels
e Shallow draft vessels
P
WC
a
b
c
d
e

DUNSTAFFNAGE MARINA

Dunstaffnage Marina Ltd
Dunbeg, by Oban, Argyll, PA37 1PX
Tel: 01631 566555 Fax: 01631 571044
Email: info@dunstaffnagemarina.com

VHF Ch M
ACCESS H24

Located just two to three miles north of Oban, Dunstaffnage Marina has been renovated to include an additional 36 fully serviced berths, a new breakwater providing shelter from NE'ly to E'ly winds and an increased amount of hard standing. Also on site is the Wide Mouthed Frog, offering a convivial bar, restaurant and accommodation with stunning views of the 13th century Dunstaffnage Castle.

The marina is perfectly placed to explore Scotland's west coast and Hebridean Islands. Only 10M NE up Loch Linnhe is Port Appin, while sailing 15M S, down the Firth of Lorne, brings you to Puldohran where you can walk to an ancient hostelry situated next to the C18 Bridge Over the Atlantic.

FACILITIES AT A GLANCE

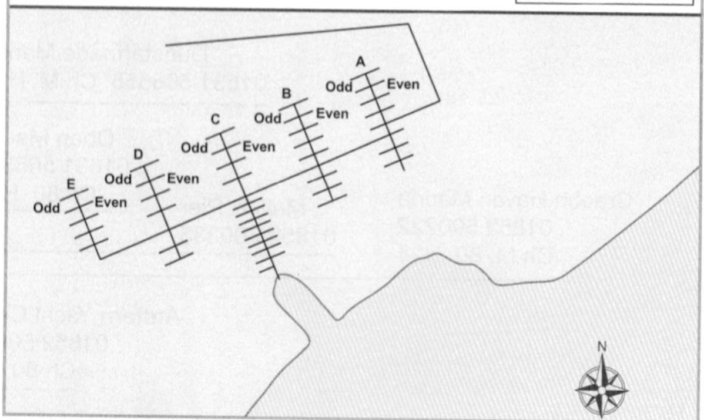
A
Odd
Even
B
Odd
Even
C
Odd
Even
D
Odd
Even
E
Odd
Even
N

OBAN MARINA

Oban Marina & Yacht Services Ltd
Isle of Kerrera, Oban, Argyll, PA34 4SX
Tel: 01631 565333
Email: info@obanmarina.com www.obanmarina.com

VHF Ch 80
ACCESS H24

Oban Marina is perfectly situated at the gateway to the Western Isles on the picturesque Isle of Kerrera. In sight of the town of Oban, it is a well serviced and popular marina offering access at all tides.

With 100 pontoons berths and 30 moorings, easily accessible diesel fuel berth, free wifi, shower block and laundry, this friendly, small marina offers visitors a warm welcome. A complementary ferry service runs to and from Oban – must be pre-booked, see website for details – where all the major facilities including restaurants, chandlery, banks and transport links are available.

FACILITIES AT A GLANCE
WC

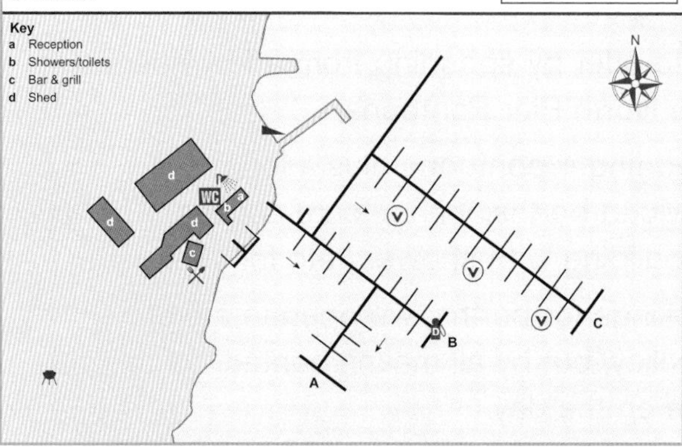
Key
a Reception
b Showers/toilets
c Bar & grill
d Shed
N
WC
A
B
C

MELFORT PIER AND HARBOUR

Melfort Pier and Harbour
Kilmelford, by Oban, Argyll
Tel: 01852 200333
Email: melharbour@aol.com www.mellowmelfort.com

VHF
ACCESS

Melfort Pier & Harbour is situated on the shores of Loch Melfort, one of the most peaceful lochs on the south west coast of Scotland. Overlooked by the Pass of Melfort and the Braes of Lorn, it lies approximately 18 miles north of Lochgilphead and 16 miles south of Oban. Its onsite facilities include showers, laundry, free Wi-Fi access and parking – pets welcome. Fuel, power and water are available at nearby Kilmelford Yacht Haven.

For those who want a few nights on dry land, Melfort Pier & Harbour offers lochside houses, each one equipped with a sauna, spa bath and balcony offering stunning views over the loch - available per night.

FACILITIES AT A GLANCE

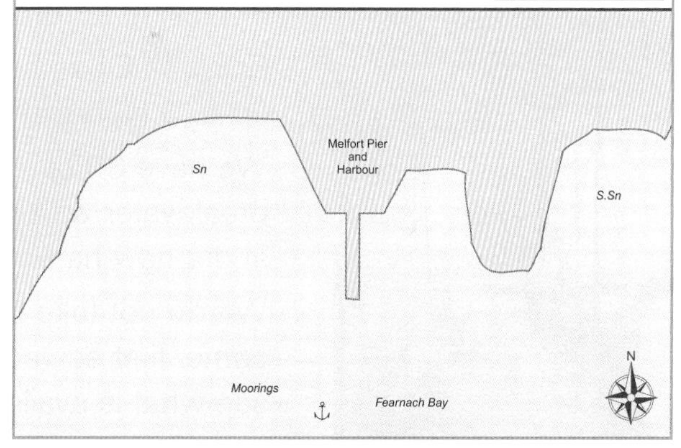
Melfort Pier and Harbour
Sn
S.Sn
Moorings
Fearnach Bay
N

CRAOBH MARINA

Craobh Marina
By Lochgilphead, Argyll, Scotland, PA31 8UA
Tel: 01852 500222 Fax: 01852 500252
Out of hours: 07702 517038
Email: info@craobhmarina.co.uk www.craobhmarina.co.uk

VHF Ch 80
ACCESS H24

Craobh Marina is idyllically situated in the heart of Scotland's most sought after cruising grounds. Not only does Craobh offer ready access to a wonderful choice of scenic cruising throughout the western isles, the marina is conveniently close to Glasgow and its international transport hub.

Craobh Marina has been developed from a near perfect natural harbour, offering secure and sheltered berthing for up to 250 vessels to 40m LOA and with a draft of 4m. With an unusually deep and wide entrance Craobh Marina provides shelter and a warm welcome for all types of craft.

FACILITIES AT A GLANCE

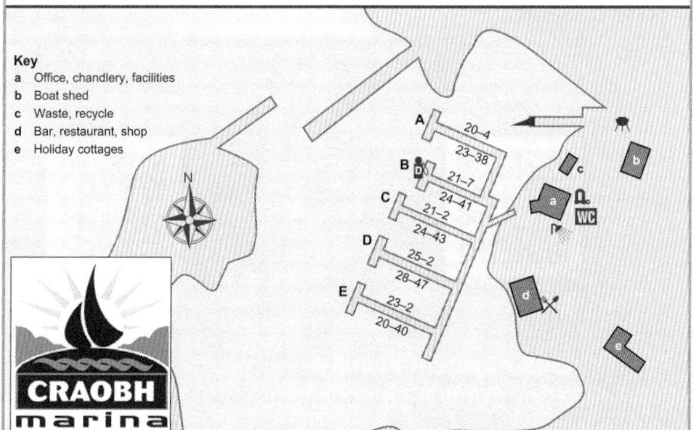
Key
a Office, chandlery, facilities
b Boat shed
c Waste, recycle
d Bar, restaurant, shop
e Holiday cottages
N
A 20–4
B 23–38
21–7
C 24–41
21–2
24–43
D 25–2
28–47
E 23–2
20–40
CRAOBH
marina

ARDFERN YACHT CENTRE

Ardfern Yacht Centre
Ardfern, by Lochgilphead, Argyll, PA31 8QN
Tel: 01852 500247 Fax: 01852 500624
www.ardfernyacht.co.uk Email: office@ardfernyacht.co.uk

VHF Ch 80
ACCESS H24

Developed around an old pier once frequented by steamers, Ardfern Yacht Centre lies at the head of Loch Craignish, one of Scotland's most sheltered and picturesque sea lochs. With several islands and protected anchorages nearby, Ardfern is an ideal place from which to cruise the west coast of Scotland and the Outer Hebrides.

The Yacht Centre comprises pontoon berths and swinging moorings as well as a workshop, boat storage and well-stocked chandlery, while a grocery store and eating places can be found in the village. Among the onshore activities available locally are horse riding, cycling, and walking.

FACILITIES AT A GLANCE

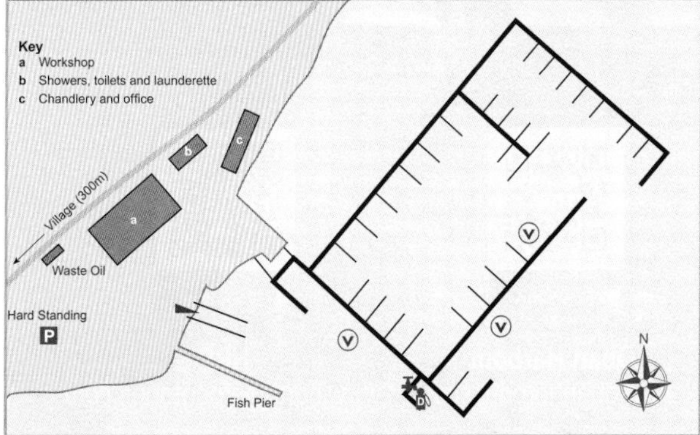
Key
a Workshop
b Showers, toilets and launderette
c Chandlery and office
Village (300m)
Waste Oil
Hard Standing
Fish Pier
N

8

SOUTH WEST SCOTLAND – Crinan Canal to Mull of Galloway

Key to Marina Plans symbols

Bottled gas	Parking
Chandler	Pub/Restaurant
Disabled facilities	Pump out
Electrical supply	Rigging service
Electrical repairs	Sail repairs
Engine repairs	Shipwright
First Aid	Shop/Supermarket
Fresh Water	Showers
Fuel - Diesel	Slipway
Fuel - Petrol	Toilets
Hardstanding/boatyard	Telephone
Internet Café	Trolleys
Laundry facilities	Visitors berths
Lift-out facilities	Wi-Fi

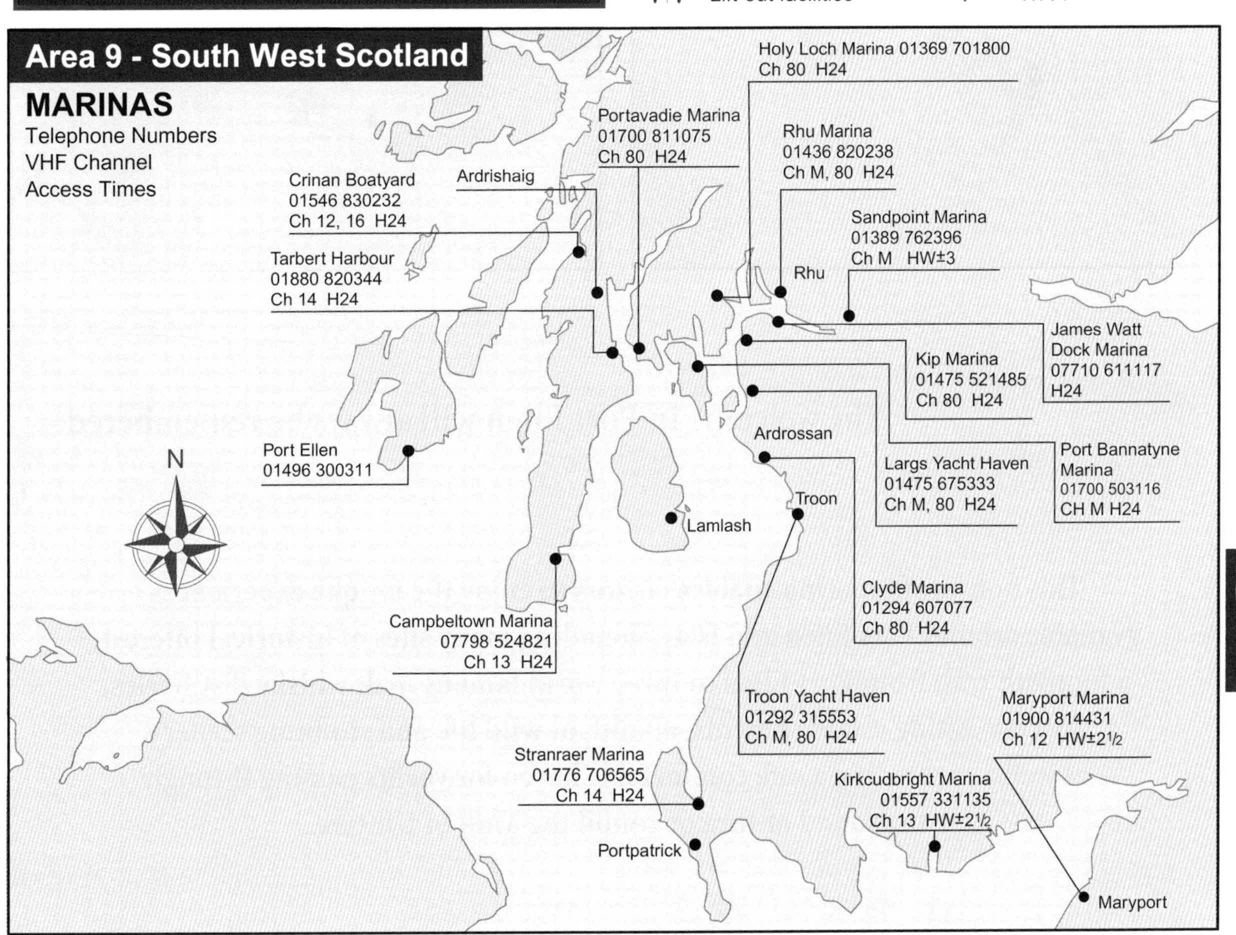
Area 9 - South West Scotland
MARINAS
Telephone Numbers
VHF Channel
Access Times
Holy Loch Marina 01369 701800
Ch 80 H24
Portavadie Marina
01700 811075
Ch 80 H24
Rhu Marina
01436 820238
Ch M, 80 H24
Crinan Boatyard
01546 830232
Ch 12, 16 H24
Ardrishaig
Sandpoint Marina
01389 762396
Ch M HW±3
Rhu
Tarbert Harbour
01880 820344
Ch 14 H24
James Watt
Dock Marina
07710 611117
H24
Kip Marina
01475 521485
Ch 80 H24
Ardrossan
N
Port Ellen
01496 300311
Largs Yacht Haven
01475 675333
Ch M, 80 H24
Port Bannatyne
Marina
01700 503116
CH M H24
Troon
Lamlash
Clyde Marina
01294 607077
Ch 80 H24
Campbeltown Marina
07798 524821
Ch 13 H24
Troon Yacht Haven
01292 315553
Ch M, 80 H24
Maryport Marina
01900 814431
Ch 12 HW±2½
Stranraer Marina
01776 706565
Ch 14 H24
Kirkcudbright Marina
01557 331135
Ch 13 HW±2½
Portpatrick
Maryport

PORT ELLEN MARINA

Port Ellen Marina
Port Ellen, Islay, Argyll, PA42 7DB
Tel: 07464 151200 www.portellenmarina.co.uk
Email: portellenmarina@outlook.com

VHF
ACCESS H24

A safe and relaxed mooring superbly located in the quaint village of Port Ellen, making it an ideal stopping off point whether planning to sail north or south. Islay is world-renowned for its many malt whisky distilleries, nine at the last count and one to reopen shortly.

Meeting guests and short term storage is trouble free with the excellent air and ferry services to the mainland and Glasgow. Once on Islay you will be tempted to extend your stay so be warned, check www.portellenmarina.com for the many reasons to visit.

FACILITIES AT A GLANCE

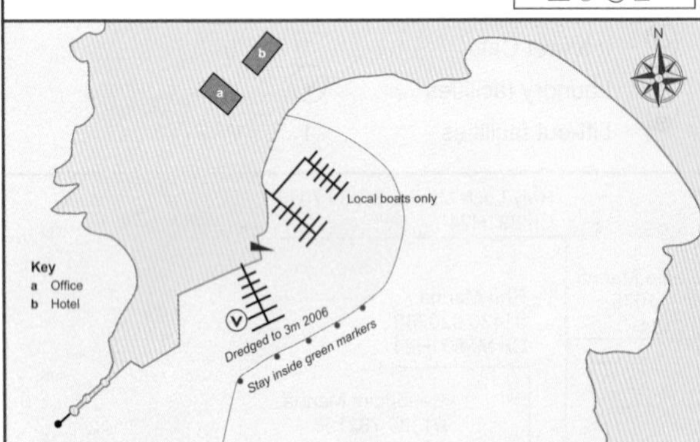
N
Local boats only
Key
a Office
b Hotel
Dredged to 3m 2006
Stay inside green markers

CRINAN BOATYARD

Crinan Boatyard Ltd
Crinan, Lochgilphead, Argyll, PA31 8SW
Tel: 01546 830232 Fax: 01546 830281
Email: info@crinanboatyard.co.uk
www.crinanboatyard.co.uk

VHF Ch 12, 16
ACCESS H24

Situated at the westerly entrance of the scenic Crinan Canal, Crinan Boatyard offers swinging moorings nightly or longer term, a fuelling/loading berth, a well stocked Chandlery, heads, showers, laundry and an experienced work force for repair work all on site. A hotel and coffee shop, just a short walk away at the Canal basin, great walking and the historic Kilmartin Glen close by are some of the attractions on shore.

The nearby town of Lochgilphead 7 miles away offers shopping and good travel links to Glasgow (85 miles) and its International Airport.

FACILITIES AT A GLANCE

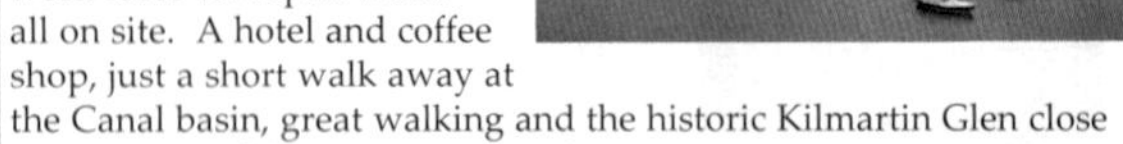

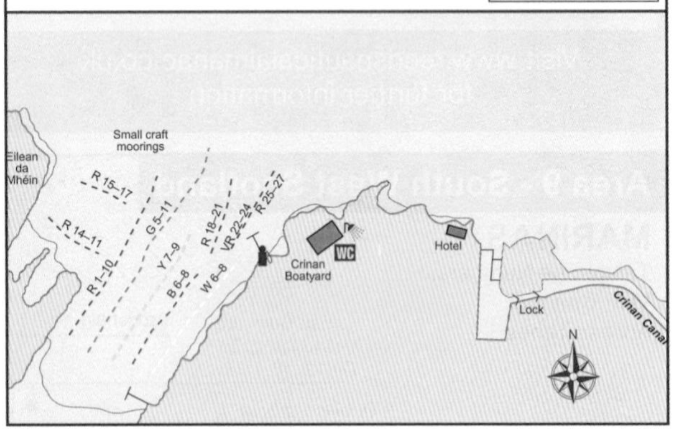
Small craft moorings
Eilean da Mhéin
R 15–17
R 14–11
R 1–10
G 5–11
Y 7–9
B 6–8
W 6–8
R 18–21
YR 22–24
R 25–27
Crinan Boatyard
WC
Hotel
Lock
Crinan Canal
N

TARBERT HARBOUR

Tarbert Harbour Authority
Harbour Office, Garval Road, Tarbert, Argyll, PA29 6TR
Tel: 01880 820344
Email: info@tarbertharbour.co.uk

VHF	Ch 14
ACCESS	H24

East Loch Tarbert is situated on the western shores of Loch Fyne. The naturally sheltered harbour is accessible through an easily navigated narrow entrance, and is a prefect stopping point for those heading north to the Crinan Canal.

The pontoons can accommodate up to 100 visiting vessels of various sizes, with fresh water, electricity and wi-fi available FOC. Toilet, shower and laundry facilities are accessible 24/7, and the unique recreation area and community marquee are available to use - perfect for families, gatherings and musters. The marina pontoons are situated at the heart of the heritage village of Tarbert, which boasts a busy festival calendar and offers a wide range of amenities.

FACILITIES AT A GLANCE

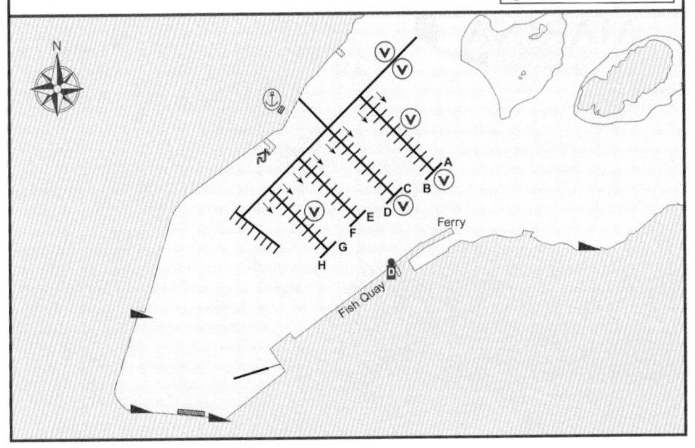
N
A
B
C
D
E
F
G
H
Ferry
Fish Quay

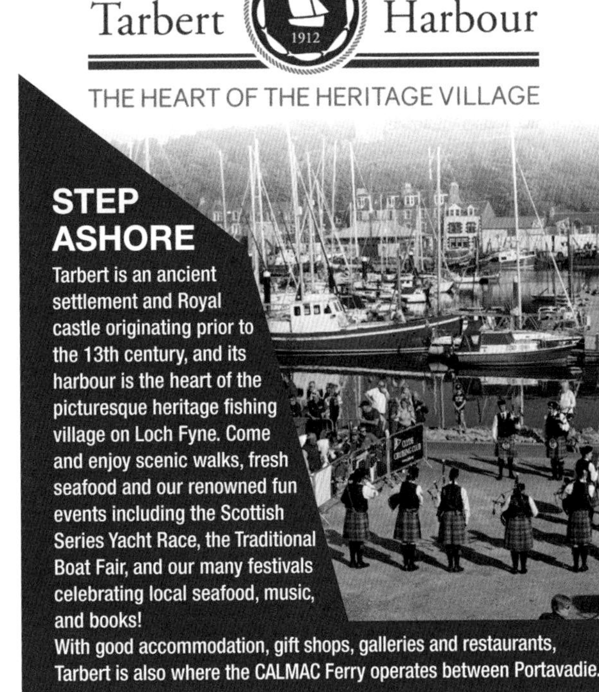
Tarbert Harbour
1912
THE HEART OF THE HERITAGE VILLAGE
STEP ASHORE
Tarbert is an ancient settlement and Royal castle originating prior to the 13th century, and its harbour is the heart of the picturesque heritage fishing village on Loch Fyne. Come and enjoy scenic walks, fresh seafood and our renowned fun events including the Scottish Series Yacht Race, the Traditional Boat Fair, and our many festivals celebrating local seafood, music, and books!
With good accommodation, gift shops, galleries and restaurants, Tarbert is also where the CALMAC Ferry operates between Portavadie.

PORT BANNATYNE MARINA

Port Bannatyne Marina
Marine Road, Port Bannatyne, PA20 0LT
Tel: 01700 503116 Mobile: 07711 319992
Email: portbannatynemarina@btconnect.com

VHF	Ch M1
ACCESS	H24

Nestled in the bay at Port Bannatyne on the Isle of Bute, the marina is set in breathtakingly beautiful surroundings. The shore facilities include toilets and showers, lifting and winter storage and all boat repairs. Free wi-fi is available throughout the marina. Protected by a breakwater and accessible H24 the marina is dredged to –2.4m CD.

The village of Port Bannatyne offers a Post Office for essential groceries. There are frequent bus services to both Rothesay and Ettrick Bay where a walk along a beautiful beach with amazing views can be completed with either a meal or tea and cake at the beach side restaurant.

FACILITIES AT A GLANCE

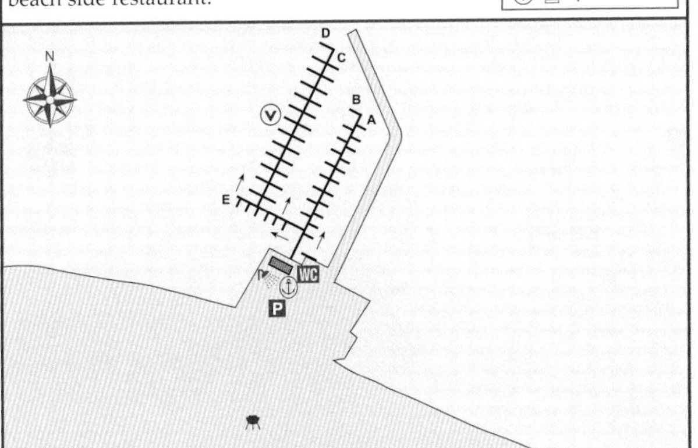
N
A
B
C
D
E

PORTAVADIE MARINA

Portavadie Marina
Portavadie, Loch Fyne, Argyll, PA21 2DA
Tel: 01700 811075 Fax: 01700 811074
Email: info@portavadiemarina.com

VHF	Ch 80
ACCESS	H24

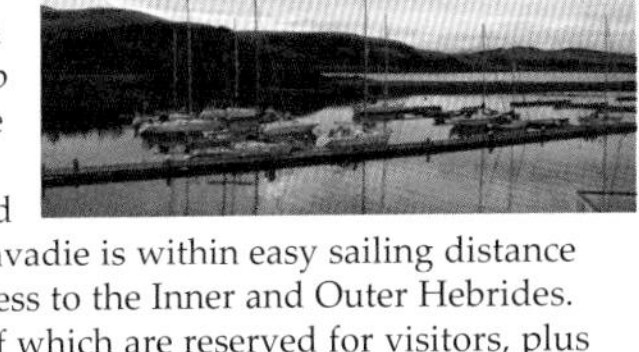

Portavadie Marina offers deep and sheltered berthing to residential and visiting boats in an area renowned for its superb cruising waters. Situated on the east side of Loch Fyne in close proximity to several islands and the famous Kyles of Bute, Portavadie is within easy sailing distance of the Crinan Canal, giving access to the Inner and Outer Hebrides. The marina has 230 berths 60 of which are reserved for visitors, plus comprehensive on shore facilities, including a choice of restaurants, bars and self catering accommodation. There is also a shop and small chandlery overlooking the marina, a dedicated fuel berth for petrol and diesel and bike hire.

This unspoiled area of Argyll which is less than two hours by road from Glasgow offers an ideal base for boat owners looking for a safe and secure haven.

FACILITIES AT A GLANCE

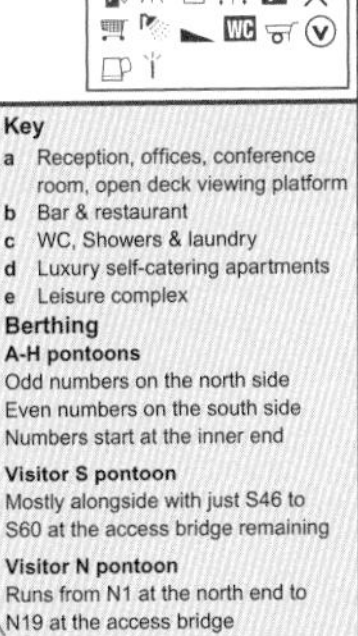
Key
a Reception, offices, conference room, open deck viewing platform
b Bar & restaurant
c WC, Showers & laundry
d Luxury self-catering apartments
e Leisure complex
Berthing
A-H pontoons
Odd numbers on the north side
Even numbers on the south side
Numbers start at the inner end
Visitor S pontoon
Mostly alongside with just S46 to S60 at the access bridge remaining
Visitor N pontoon
Runs from N1 at the north end to N19 at the access bridge

CAMPBELTOWN MARINA

Campbeltown Marina Ltd
Dubh Artach, Roading, Campbeltown, PA28 6LU
Tel: 07798 524821
Email: campbeltownmarina@btinternet.com

VHF	Ch 13
ACCESS	H24

Campbeltown Marina is a brand new facility opened in June 2015 and is situated in the town centre at the head of the deep, sheltered waters of Campbeltown Loch on the SE aspect of the Kintyre Peninsula. It is within easy reach of the Antrim Coast, Ayrshire and the Upper Clyde. Diesel is available at the Old Quay and gas is across the road. Petrol can be bought a 5-minute walk away and a well stocked chandlery is situated in the town centre.

Campbeltown is the perfect getaway destination with plenty to offer the whole family. Golf, cycling and walking routes, modern swimming pool and horse riding are some of the activities on offer. Situated directly in the town centre there is a wide choice of shops, cafes, bars, restaurants and supermarkets within easy walking distance.

FACILITIES AT A GLANCE

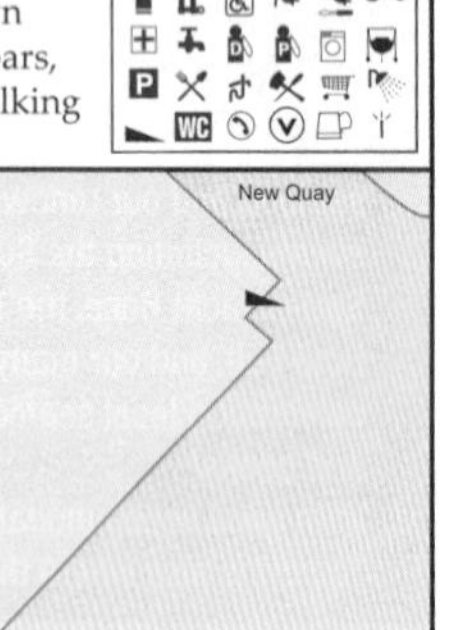
New Quay
Old Quay
Fish market
Concrete breakwater
Lifeboat station
WC
Tourist info
Small boats

HOLY LOCH MARINA

Holy Loch Marina
Rankin's Brae, Sandbank, Dunoon, PA23 8FE
Tel: 01369 701800
Email: info@holylochmarina.co.uk www.holylochmarina.co.uk

VHF	Ch 80
ACCESS	H24

Holy Loch Marina, the marine gateway to Loch Lomond and the Trossachs National Park, lies on the south shore of the loch, roughly half a mile west of Lazaretto Point. Holy Loch is among the Clyde's most beautiful natural harbours and, besides being a peaceful location, offers an abundance of wildlife, places of local historical interest as well as excellent walking and cycling through the Argyll Forest Park. The marina can be entered in all weather conditions and is within easy sailing distance of Loch Long and Upper Firth.

FACILITIES AT A GLANCE

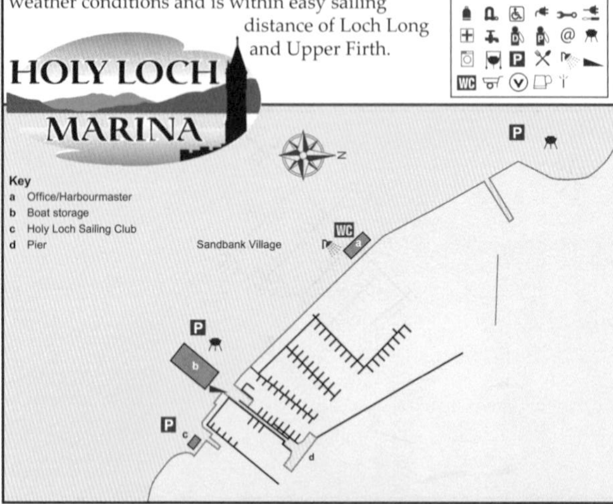
HOLY LOCH
MARINA
Key
a Office/Harbourmaster
b Boat storage
c Holy Loch Sailing Club
d Pier
Sandbank Village
WC
P

RHU MARINA

Rhu Marina
Boatfolk Marinas Ltd, Pier Road, Rhu, G84 8LH
Tel: 01436 820238
Email: rhu@boatfolk.co.uk
www.boatfolk.co.uk/rhumarina

VHF	Ch M, 80
ACCESS	H24

Located on the north shore of the Clyde Estuary, Rhu Marina is accessible at all states of the tide and can accommodate yachts up to 24m in length. It also operates 40 swinging moorings in the bay adjacent to the marina, with a ferry service provided.

Within easy walking distance of the marina is Rhu village, a conservation village incorporating a few shops, a pub and the beautiful Glenarn Gardens as well as the Royal Northern & Clyde Yacht Club. A mile or two to the east lies the holiday town of Helensburgh, renowned for its attractive architecture and elegant parks and gardens, while Glasgow city is just 25 miles away and can be easily reached by train.

FACILITIES AT A GLANCE

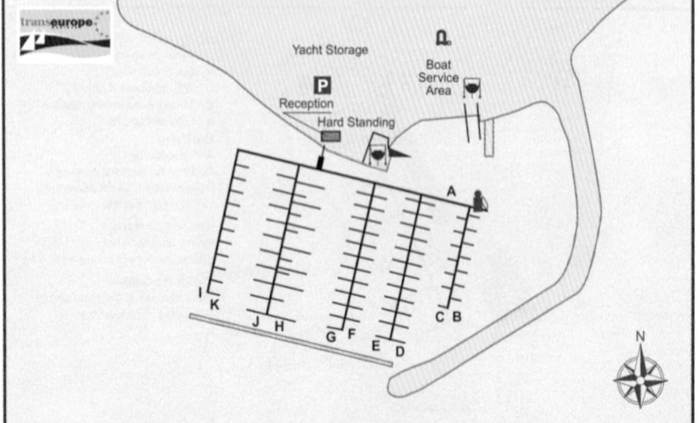
transeurope
Yacht Storage
Boat Service Area
Reception
Hard Standing
P
A
B
C
D
E
F
G
H
J
K
I
N

SANDPOINT MARINA

Sandpoint Marina Ltd
Sandpoint, Woodyard Road, Dumbarton, G82 4BG
Tel: 01389 762396 Fax: 01389 732605
Email: sales@sandpoint-marina.co.uk
www.sandpoint-marina.co.uk

VHF	
ACCESS	HW±3

Lying on the north bank of the Clyde estuary on the opposite side of the River Leven from Dumbarton Castle, Sandpoint Marina provides easy access to some of the most stunning cruising grounds in the United Kingdom. It is an independently run marina, offering a professional yet personal service to every boat owner. Among the facilities to hand are an on site chandlery, storage areas, a 40 ton travel hoist and 20 individual workshop units.

Within a 20-minute drive of Glasgow city centre, the marina is situated close to the shores of Loch Lomond, the largest fresh water loch in Britain.

FACILITIES AT A GLANCE

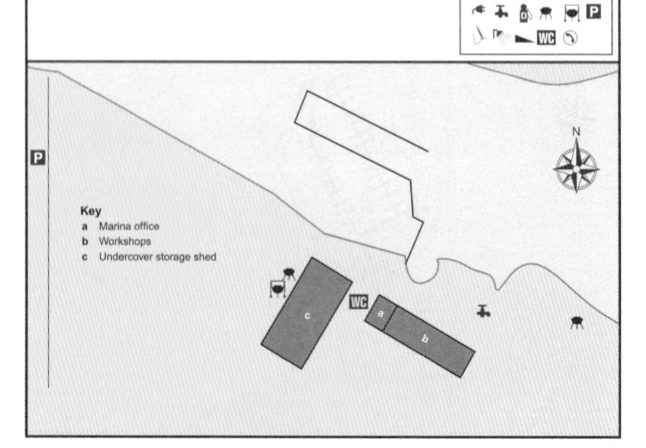
Key
a Marina office
b Workshops
c Undercover storage shed
P
WC
N

KIP MARINA

Kip Marina, The Yacht Harbour
Inverkip, Renfrewshire, Scotland, PA16 0AS
Tel: 01475 521485
www.kipmarina.co.uk Email: info@kipmarina.co.uk

VHF	Ch 80
ACCESS	H24

Inverkip is a small village which lies on the south shores of the River Kip as it enters the Firth of Clyde. Once established for fishing, smuggling and, in the 17th century, witch-hunts, it became a seaside resort in the 1860s as a result of the installation of the railway. Today it is a yachting centre, boasting a state-of-the-art marina with over 600 berths and full boatyard facilities. With the capacity to accommodate yachts of up to 23m LOA, Kip Marina offers direct road and rail access to Glasgow and its international airport, therefore making it an ideal location for either a winter lay up or crew changeover.

FACILITIES AT A GLANCE

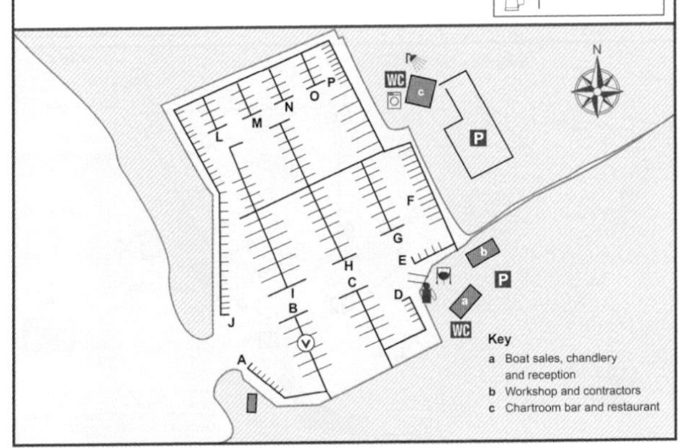
N
A
B
C
D
E
F
G
H
I
J
L
M
N
O
P
Key
a Boat sales, chandlery and reception
b Workshop and contractors
c Chartroom bar and restaurant

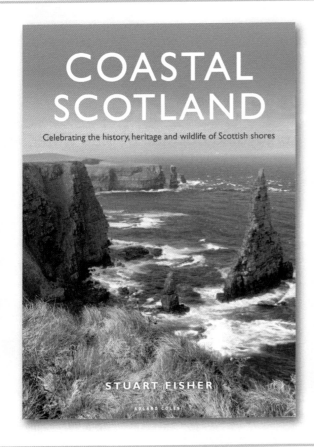

JAMES WATT DOCK MARINA

James Watt Dock Marina
East Hamilton Street
Greenock, Renfrewshire, PA15 2TD
Tel: 01475 729838
www.jwdmarina.co.uk Email: info@jwdmarina.co.uk

VHF	80
ACCESS	H24

James Watt Dock Marina offers around 170 berths alongside or on finger pontoons for craft ranging in size from 7m to 100+m within a historic dock setting. With excellent motorway and public transport connections and easy access to some of the best sailing waters at all states of tide, James Watt Dock Marina provides unbeatable opportunities for boaters

The marina is a short distance from Greenock's cinema, pool, ice rink, restaurants and shops, and with nearby transport connections, the marina will be a great location for both visitors and regular berthers.

FACILITIES AT A GLANCE

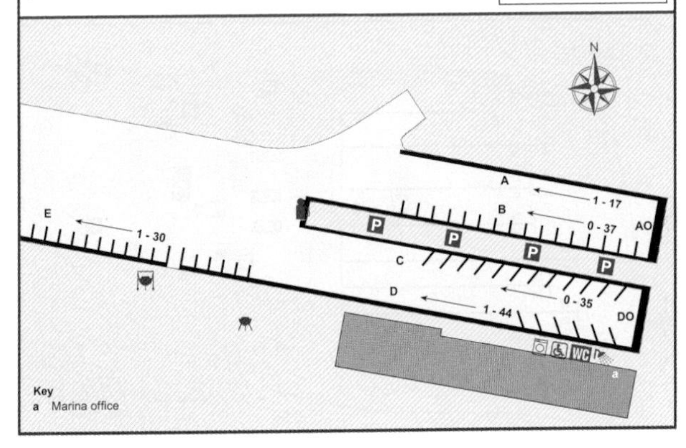
N
A
1 - 17
B
0 - 37
AO
C
D
0 - 35
1 - 44
DO
E
1 - 30
Key
a Marina office

LARGS YACHT HAVEN

Largs Yacht Haven Ltd
Irvine Road, Largs, Ayrshire, KA30 8EZ
Tel: 01475 675333
Email: largs@yachthavens.com www.yachthavens.com

VHF	Ch M, 80
ACCESS	H24

Largs Yacht Haven offers a superb location among lochs and islands, with numerous fishing villages and harbours nearby. Sheltered cruising can be enjoyed in the inner Clyde, while the west coast and Ireland are only a day's sail away. With a stunning backdrop of the Scottish mountains, Largs incorporates 700 fully serviced berths and provides a range of on site facilities including chandlers, sailmakers, engineers, shops, restaurants and club.

A 20-minute coastal walk brings you to the town of Largs, which has all the usual amenities as well as excellent road and rail connections to Glasgow.

FACILITIES AT A GLANCE

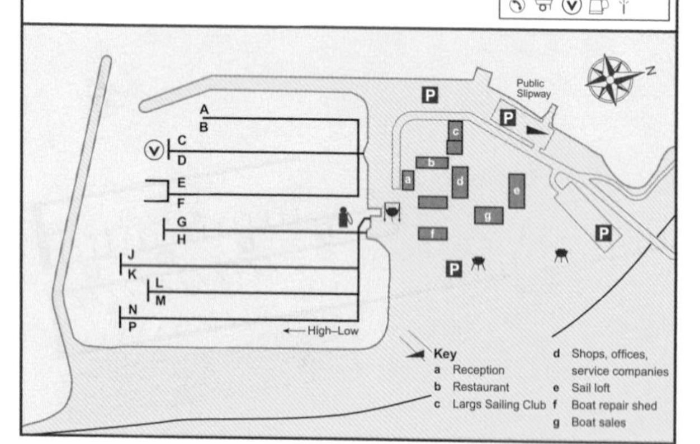
A
B
C
D
E
F
G
H
J
K
L
M
N
P
High–Low
Public Slipway
Key
a Reception
b Restaurant
c Largs Sailing Club
d Shops, offices, service companies
e Sail loft
f Boat repair shed
g Boat sales

CLYDE MARINA

Clyde Marina Ltd
The Harbour, Ardrossan, Ayrshire, KA22 8DB
Tel: 01294 607077 Fax: 01294 607076
www.clydemarina.com Email: info@clydemarina.com

VHF	Ch 80
ACCESS	H24

CLYDE MARINA

Situated on the Clyde Coast between Irvine and Largs, Clyde Marina is Scotland's third largest marina and boatyard. It is set in a landscaped environment boasting a 50T hoist and active boat sales. A deep draft marina berthing vessels up to 30m LOA, draft up to 5m. Peviously accommodated vessels include tall ships and Whitbread 60s plus a variety of sail and power craft. Fully serviced pontoons plus all the yard facilities you would expect from a leading marina including boatyard and boatshed for repairs or storage. Good road and rail connections and only 30 minutes from Glasgow and Prestwick airports.

FACILITIES AT A GLANCE

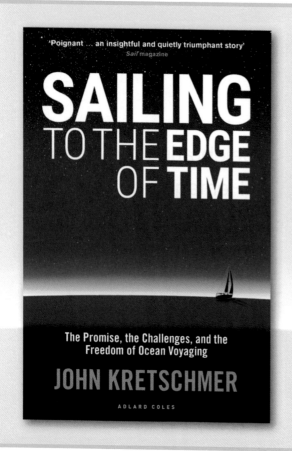

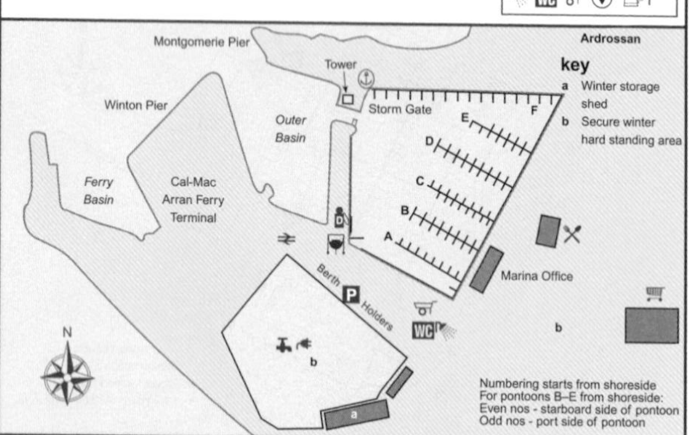
Montgomerie Pier
Ardrossan
Tower
Storm Gate
Winton Pier
Outer Basin
Ferry Basin
Cal-Mac Arran Ferry Terminal
Berth Holders
Marina Office
A
B
C
D
E
F
N
key
a Winter storage shed
b Secure winter hard standing area
Numbering starts from shoreside
For pontoons B–E from shoreside:
Even nos - starboard side of pontoon
Odd nos - port side of pontoon

TROON YACHT HAVEN

Troon Yacht Haven Ltd
The Harbour, Troon, Ayrshire, KA10 6DJ
Tel: 01292 315553
Email: troon@yachthavens.com
www.yachthavens.com

VHF Ch 80
ACCESS H24

Troon Yacht Haven, situated on the Southern Clyde Estuary, benefits from deep water at all states of the tide. Tucked away in the harbour of Troon, it is well sheltered and within easy access of the town centre.

There are plenty of cruising opportunities to be had from here, whether it be hopping across to the Isle of Arran, with its peaceful anchorages and mountain walks, sailing round the Mull or through the Crinan Canal to the Western Isles, or heading for the sheltered waters of the Clyde.

FACILITIES AT A GLANCE
WC

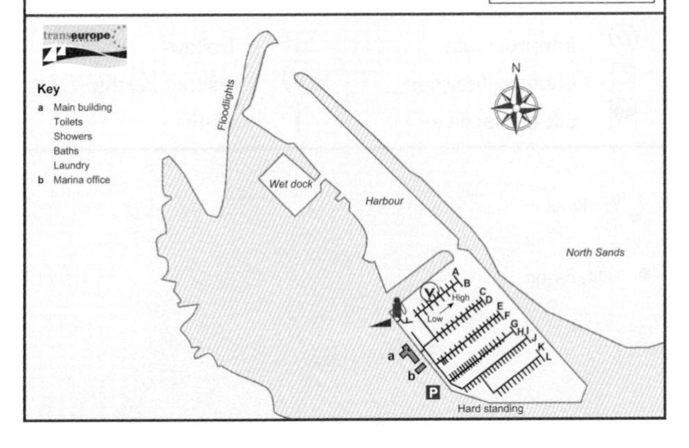
transeurope
Key
a Main building
Toilets
Showers
Baths
Laundry
b Marina office
Floodlights
N
Wet dock
Harbour
North Sands
High
Low
A
B
C
D
E
F
G
H
J
K
L
a
b
P
Hard standing

STRANRAER MARINA

Stranraer Marina
West Pier, Market Street, Stranraer, DG9 7RE
Tel: 01776 706565 Mob: 07734 073421
Email: lesley.smith@dumgal.gov.uk

VHF Ch 14
ACCESS H24

Stranraer Marina is at the southern end of beautiful Loch Ryan with 24H access and modern facilities with a 30T boat crane, transporter and hardstanding available. The visitor berths are those immediately adjacent to the breakwater. Craft over 12m are advised to call ahead for availability.

Stranraer town centre is only a short walk from the marina. Berthing rates are available at www.dumgal.gov.uk/harbours. It should be noted that the marina is exposed in strong N winds. Two ferry terminals are located on the E side of the loch approximately 6M N of the marina and extra care should be taken in this area.

FACILITIES AT A GLANCE
WC

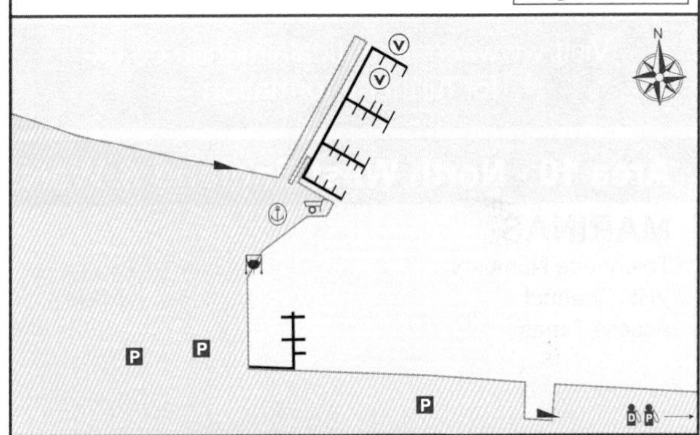
N
P
P
P

KIRKCUDBRIGHT MARINA

Kirkcudbright Marina
Militia House, English Street, Dumfries, DG1 2HR
Tel: 01557 331135 Mob: 07709 479663
Email: kirkcudbright.harbour@dumgal.gov.uk

VHF Ch 16, 73
ACCESS HW±2.5

A very well sheltered picturesque marina accessed HW+/-2.5 hrs via a 3.5 mile long, narrow channel that is well marked and lit, contact should be made with Range Safety vessel 'Gallovidian' prior to approach. Limited visitors berths so vessels should contact the harbour master in advance.

The marina is only 250m from the centre of Kirkcudbright, an historic 'artist's' town where visitors may enjoy a wide range of facilities and tourist attractions including castle, museum, tollbooth, art galleries and traditional shops. There is a superb programme of summer festivities.

FACILITIES AT A GLANCE
WC

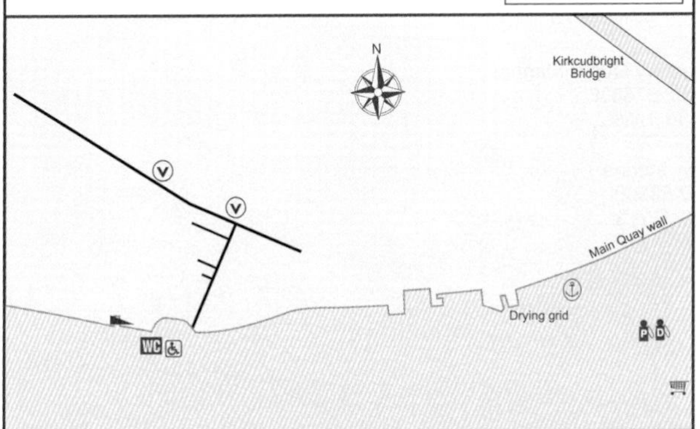
N
Kirkcudbright Bridge
Main Quay wall
Drying grid
WC

MARYPORT MARINA

Maryport Harbour and Matine Ltd
Marine Road, Maryport, Cumbria, CA15 8AY
Tel: 01900 814431
www.maryportmarina.com
Email: enquires@maryportmarina.com

VHF Ch 12,16
ACCESS HW±2.5

Maryport Marina is located in the historic Senhouse Dock, which was originally built for sailing clippers in the late 19th century. The old stone harbour walls provide good shelter to the 190 berths from the prevailing south westerlies.

Maryport town centre and its shops, pubs and other amenities is within easy walking distance from the marina. Maryport a perfect location from which to explore the west coast of Scotland as well as the Isle of Man and the Galloway Coast. For those who wish to venture inland, then the Lake District is only seven miles away.

FACILITIES AT A GLANCE
WC

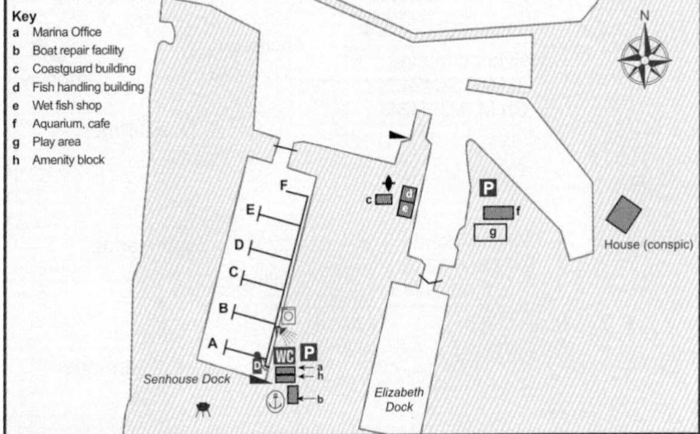
Key
a Marina Office
b Boat repair facility
c Coastguard building
d Fish handling building
e Wet fish shop
f Aquarium, cafe
g Play area
h Amenity block
N
A
B
C
D
E
F
House (conspic)
Senhouse Dock
Elizabeth Dock
WC
P

NW ENGLAND, ISLE OF MAN & N WALES – Mull of Galloway to Bardsey Is

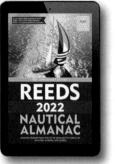

Key to Marina Plans symbols

- Bottled gas
- Chandler
- Disabled facilities
- Electrical supply
- Electrical repairs
- Engine repairs
- First Aid
- Fresh Water
- Fuel - Diesel
- Fuel - Petrol
- Hardstanding/boatyard
- Internet Café
- Laundry facilities
- Lift-out facilities
- Parking
- Pub/Restaurant
- Pump out
- Rigging service
- Sail repairs
- Shipwright
- Shop/Supermarket
- Showers
- Slipway
- Toilets
- Telephone
- Trolleys
- Visitors berths
- Wi-Fi

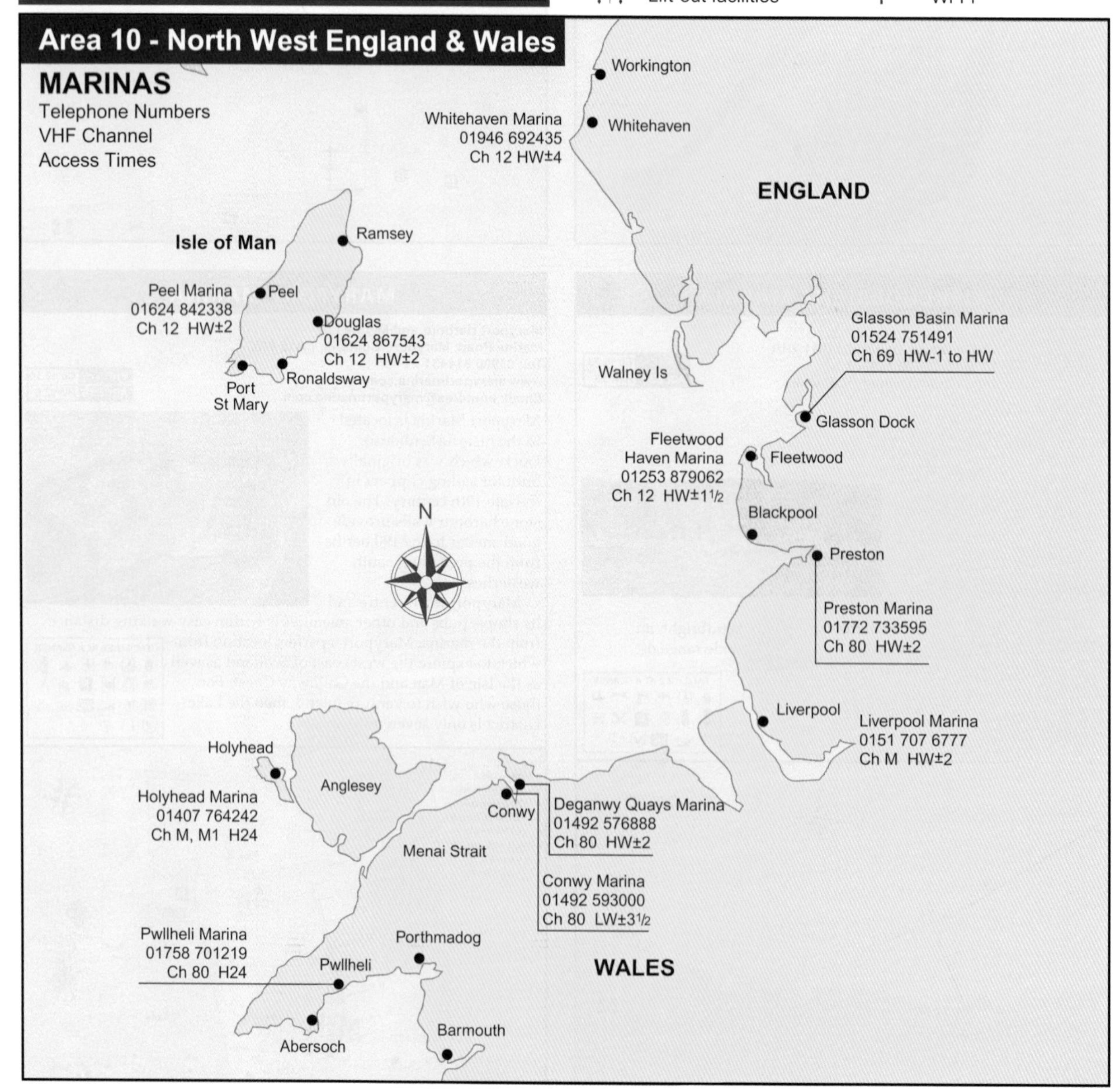
Area 10 - North West England & Wales
MARINAS
Telephone Numbers
VHF Channel
Access Times
Workington
Whitehaven Marina
01946 692435
Ch 12 HW±4
Whitehaven
ENGLAND
Isle of Man
Ramsey
Peel Marina
01624 842338
Ch 12 HW±2
Peel
Douglas
01624 867543
Ch 12 HW±2
Ronaldsway
Port St Mary
Walney Is
Glasson Basin Marina
01524 751491
Ch 69 HW-1 to HW
Glasson Dock
Fleetwood Haven Marina
01253 879062
Ch 12 HW±1½
Fleetwood
Blackpool
Preston
N
Preston Marina
01772 733595
Ch 80 HW±2
Liverpool
Liverpool Marina
0151 707 6777
Ch M HW±2
Holyhead
Anglesey
Holyhead Marina
01407 764242
Ch M, M1 H24
Conwy
Deganwy Quays Marina
01492 576888
Ch 80 HW±2
Menai Strait
Conwy Marina
01492 593000
Ch 80 LW±3½
Porthmadog
Pwllheli Marina
01758 701219
Ch 80 H24
Pwllheli
WALES
Barmouth
Abersoch

WHITEHAVEN MARINA

Whitehaven Marina Ltd
Harbour Office, Bulwark Quay, Whitehaven, Cumbria, CA28 7HS
Tel: 01946 692435
Email: enquiries@whitehavenmarina.co.uk
www.whitehavenmarina.co.uk

VHF	Ch 12
ACCESS	HW±4

Whitehaven Marina can be found at the south-western entrance to the Solway Firth, providing a strategic departure point for those yachts heading for the Isle of Man, Ireland or Southern Scotland. The harbour is one of the more accessible ports of refuge in NW England, affording a safe entry in most weathers. The approach channel across the outer harbour is dredged to about 1.0m above chart datum, allowing entry into the inner harbour via a sea lock at around HW±4. Over 100 new walk ashore berths were installed in 2013.

FACILITIES AT A GLANCE

Conveniently situated for visiting the Lake District, Whitehaven is an attractive Georgian town, renowned in the C18 for its rum and slave imports.

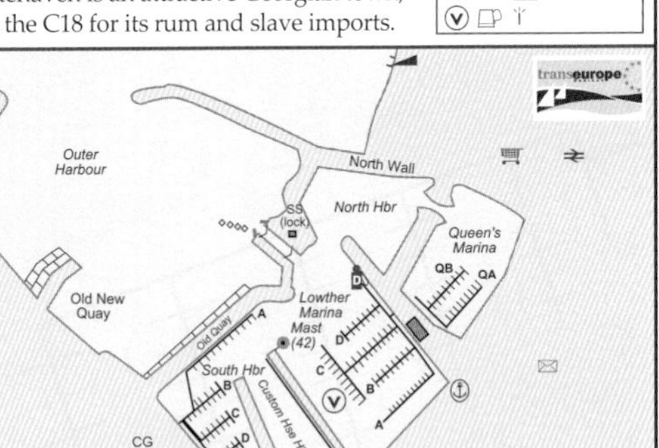
N
transeurope
Outer Harbour
North Wall
North Hbr
SS (lock)
Queen's Marina
QB
QA
Old New Quay
Lowther Marina Mast (42)
Old Quay
South Hbr
Custom Hse Hbr
CG
Heritage Centre
Lime Tongue
Sugar Tongue

GLASSON WATERSIDE & MARINA

Glasson Waterside & Marina
School Lane, Glasson Dock, Lancaster, LA2 0AW
Tel: 01524 751491 Fax: 01524 752626
Email: glasson@aquavista.com
www.aquavista.com

VHF	Ch 69
ACCESS	HW-1 to HW

Glasson Basin Marina lies on the River Lune, west of Sunderland Point. Access is via the outer dock which opens 45 minutes before HW. Liverpool and thence via BWB lock into the inner basin. It is recommended to leave Lune No. 1 Buoy approx 1.5 hrs before HW. Contact the dock on Channel 69. The Marina can only be contacted by telephone. All the necessary requirements can be found either on site or within easy reach of Glasson Dock, including boat, rigging and sail repair services as well as a launderette, ablution facilities, shops and restaurants.

FACILITIES AT A GLANCE

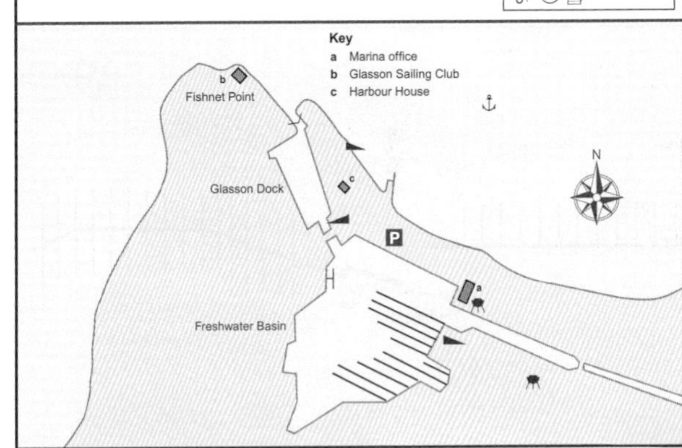
Key
a Marina office
b Glasson Sailing Club
c Harbour House
Fishnet Point
Glasson Dock
Freshwater Basin
N
P

Where the Lancaster canal meets the sea
GOLD ANCHOR
Glasson Waterside & Marina
One of our unique and special little nooks of life by the water, where you'll find:
• Gated key-fob access and CCTV because we're safe and secure as standard
• Shower, toilets and laundry facilities to make your life a little bit easier
• Well-stocked chandlery for all the bits and pieces you might need
• Serviced boatyard and workshop with a 50 tonne hoist and Glasson Marine Services for repair, paintwork and rigging.
We'll see you soon - visitors are always welcome.
Find us at: www.aquavista.com
AQUAVISTA
discover the life

DOUGLAS MARINA

Douglas Marina
Sea Terminal Building, Douglas, IM1 2RF
Tel: 01624 686627 Fax: 01624 686612
www.gov.im/harbours/
Email: enquiries.harboursdoi@gov.im

VHF	Ch 12, 16
ACCESS	HW±2

Douglas Marina is accessible HW±2 with 2.5m retained at LW. The depth of water inside the marina can vary so please advise draft. The maximum length accommodated on pontoons is 15m. Wall berths are also available. Douglas Marina Operations Centre requires clearance for entrance to the outer harbour due to commercial traffic. Please inform arrival on VHF Ch 12 10 minutes before port entry.

The marina has many facilities including electricity, fresh water, diesel, lift out, drying pad, gas, chandlery and showers/toilet facilities.

Douglas Marina is in the heart of the Isle of Man's capital so all local amenities and transport links such as the Steam Railway just a short walk.

FACILITIES AT A GLANCE

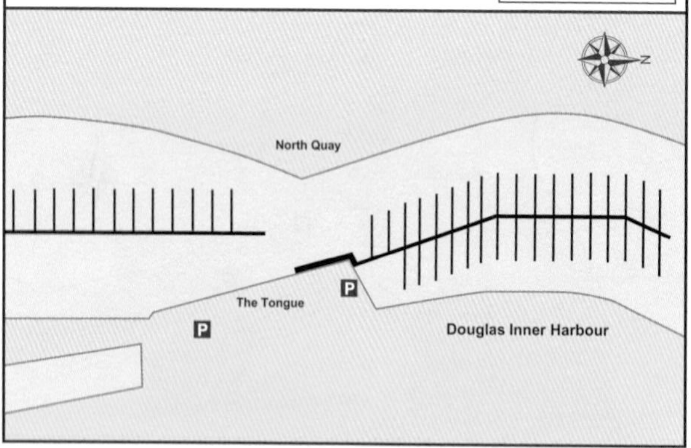
Z
North Quay
The Tongue
P
P
Douglas Inner Harbour

PEEL MARINA

Peel Marina
The Harbour Office, East Quay, Peel, IM5 1AR
Tel: 01624 842338 Fax: 01624 843610
www.gov.im/harbours/
Email: harbours@gov.im

VHF	Ch 12
ACCESS	HW±2

The Inner harbour has provision on pontoons for visiting vessels up 15m with rafting also available on the harbour walls. Access is available HW±2hrs with a maximum draft of 2.5m retained at low water but this does vary so please advise vessel dimensions on approach via VHF Ch 12.

Fresh water and electricity are available on all pontoon and some wall berths. Diesel fuel is available at the quayside with petrol sourced from a local forecourt in Peel.

Peel is a lovely active fishing port with many local amenities and a beautiful castle overlooking the harbour.

FACILITIES AT A GLANCE

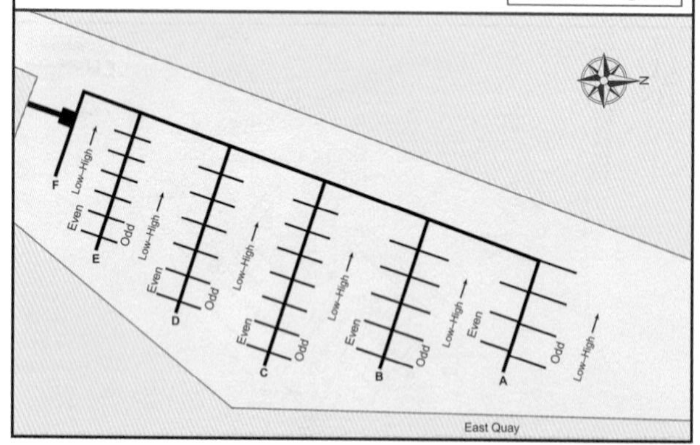
Z
F
E
D
C
B
A
Low-High
Even
Odd
East Quay

FLEETWOOD HAVEN MARINA

Fleetwood Haven Marina
c/o ABP, Port & Marina Office, Fleetwood, FY7 8BP
Tel: 01253 879062 Fax: 01253 879063
Email: fleetwoodhaven@abports.co.uk

VHF	Ch 12
ACCESS	HW±1.5

Fleetwood Haven Marina provides a good location from which to cruise Morecambe Bay and the Irish Sea. To the north west is the Isle of Man, the north is the Solway Firth and the Clyde Estuary, while to the south west is Conwy, the Menai Straits and Holyhead.

Tucked away in a protected dock which dates to 1835, the marina has 266 full service berths and offers extensive facilities including a 75-tonne boat hoist, laundry and a first class shower/bathroom block.

Call Fleetwood Dock Radio on VHF Channel 12 (Tel 01253 872351) for permission to enter the dock channel.

FACILITIES AT A GLANCE

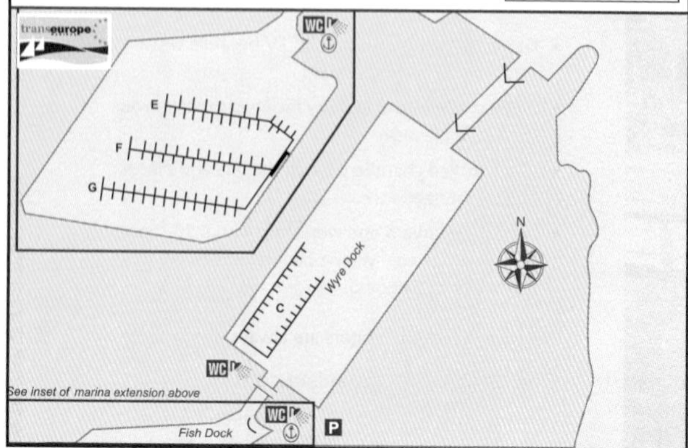
transeurope
E
F
G
N
C
Wyre Dock
See inset of marina extension above
Fish Dock
WC
P

PRESTON MARINA

Preston Marine Services Ltd
The Boathouse, Navigation Way, Preston, PR2 2YP
Tel: 01772 733595
Email: info@prestonmarina.co.uk www.prestonmarina.co.uk

VHF	Ch 80
ACCESS	HW±2

Preston Marina forms part of the comprehensive Riversway Docklands development, meeting all the demands of modern day boat owners. With the docks' history dating back over 100 years, today the marina comprises 40 acres of fully serviced pontoon berths sheltered behind the lock gates.

Lying 15 miles up the River Ribble, which itself is an interesting cruising ground with an abundance of wildlife, Preston is well placed for sailing to parts of Scotland, Ireland or Wales. The Docklands development includes a wide choice of restaurants, shops and cinemas as well as being in easy reach of all the cultural and leisure facilities provided by a large town.

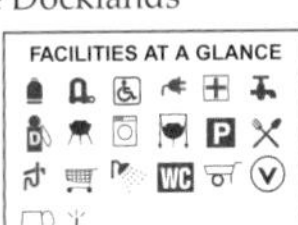
FACILITIES AT A GLANCE

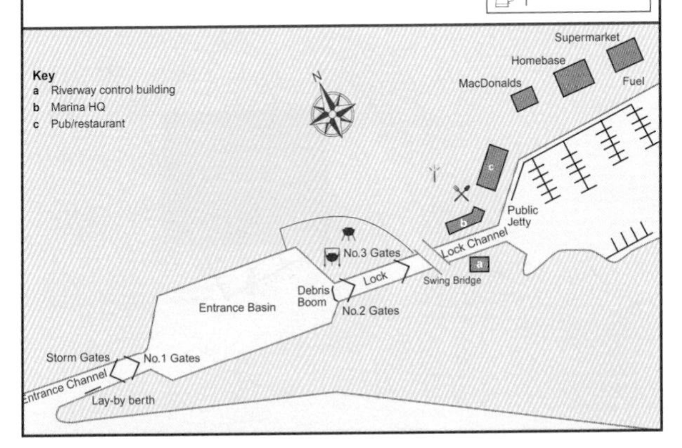
Key
a Riverway control building
b Marina HQ
c Pub/restaurant
Supermarket
Homebase
MacDonalds
Fuel
Public Jetty
Lock Channel
No.3 Gates
Lock
Swing Bridge
Debris Boom
No.2 Gates
Entrance Basin
Storm Gates
No.1 Gates
Entrance Channel
Lay-by berth

LIVERPOOL MARINA

Liverpool Marina
Coburg Wharf, Sefton Street, Liverpool, L3 4BP
Tel: 0151 707 6777 Fax: 0151 707 6770
Email: mail@liverpoolmarina.co.uk

VHF	Ch M
ACCESS	HW±2

Liverpool Marina is ideally situated for yachtsmen wishing to cruise the Irish Sea. Access is through a computerised lock that opens two and a half hours either side of high water between 0600 and 2200 daily. Once in the marina, you can enjoy the benefits of the facilities on offer, including a first class club bar and restaurant.

Liverpool is a thriving cosmopolitan city, with attractions ranging from numerous bars and restaurants to museums, art galleries and the Beatles Story.

FACILITIES AT A GLANCE

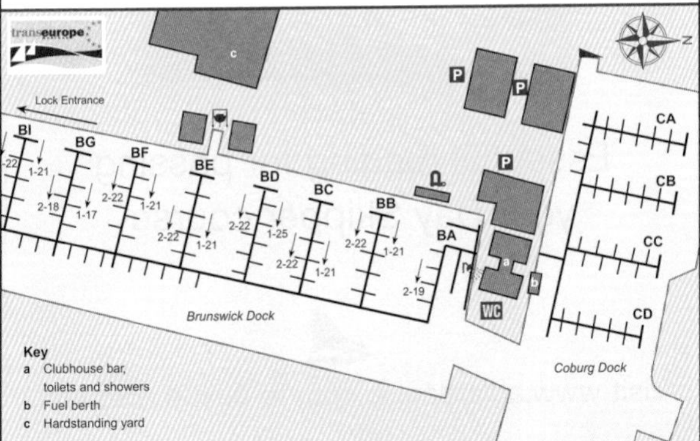
transeurope
Lock Entrance
BI
BG
BF
BE
BD
BC
BB
BA
CA
CB
CC
CD
Brunswick Dock
Coburg Dock
Key
a Clubhouse bar, toilets and showers
b Fuel berth
c Hardstanding yard

CONWY MARINA

Conwy Marina
Conwy, LL32 8EP
Tel: 01492 593000
Email: conwy@boatfolk.co.uk
www.boatfolk.co.uk/conwymarina

VHF	Ch 80
ACCESS	LW±3.5

Situated in an area of outstanding natural beauty, with the Mountains of Snowdonia National Park providing a stunning backdrop, Conwy is the first purpose-built marina to be developed on the north coast of Wales. Enjoying a unique site next to the 13th century Conwy Castle, the third of Edward I's great castles, it provides a convenient base from which to explore the cruising grounds of the North Wales coast. The unspoilt coves of Anglesey and the beautiful Menai Straits prove a popular destination, while further afield are the Llyn Peninsula and the Islands of Bardsey and Tudwells. The marina incorporates about 500 fully serviced berths which are accessible through a barrier gate at half tide.

FACILITIES AT A GLANCE

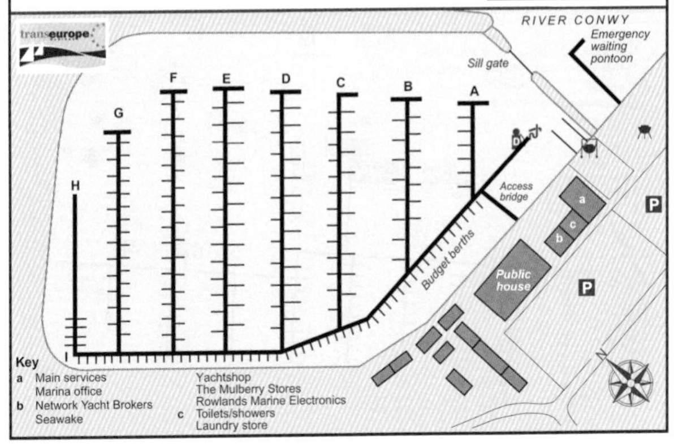
transeurope
RIVER CONWY
Emergency waiting pontoon
Sill gate
Access bridge
Budget berths
Public house
Key
a Main services
Marina office
b Network Yacht Brokers
Seawake
Yachtshop
The Mulberry Stores
Rowlands Marine Electronics
c Toilets/showers
Laundry store

DEGANWY MARINA

Deganwy Marina
Deganwy, Conwy, LL31 9DJ
Tel: 01492 576888 Fax: 01492 580066
Email: enquiries@deganwymarina.co.uk
www.deganwymarina.co.uk

VHF	Ch 80
ACCESS	HW±3

Deganwy Marina is located in the centre of the north Wales coastline on the estuary of the Conwy River and sits between the river and the small town of Deganwy with the beautiful backdrop of the Vardre hills. The views from the marina across the Conwy River are truly outstanding with the medieval walled town and Castle of Conwy outlined against the foothills of the Snowdonia National Park.

Deganwy Marina has 165 fully serviced berths, which are accessed via a tidal gate between half tide and high water.

FACILITIES AT A GLANCE

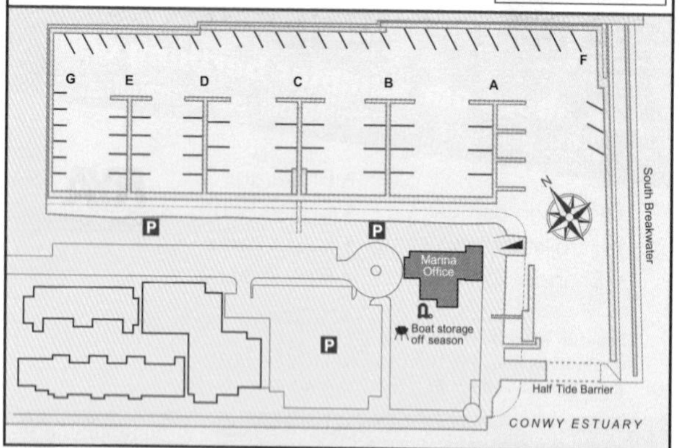
G
E
D
C
B
A
F
South Breakwater
P
Marina Office
Boat storage off season
Half Tide Barrier
CONWY ESTUARY

HOLYHEAD MARINA

Holyhead Marina Ltd
Newry Beach, Holyhead, Gwynedd, LL65 1YA
Tel: 01407 764242 Fax: 01407 769152
Email: info@holyheadmarina.co.uk

VHF	Ch M
ACCESS	H24

One of the few natural deep water harbours on the Welsh coast, Anglesey is conveniently placed as a first port of call if heading to North Wales from the North, South or West. Its marina at Holyhead, accessible at all states of the tide, is sheltered by Holyhead Mountain as well as an enormous harbour breakwater and extensive floating breakwaters, therefore offering good protection from all directions.

Anglesey boasts numerous picturesque anchorages and beaches in addition to striking views over Snowdonia, while only a tide or two away are the Isle of Man and Eire.

FACILITIES AT A GLANCE

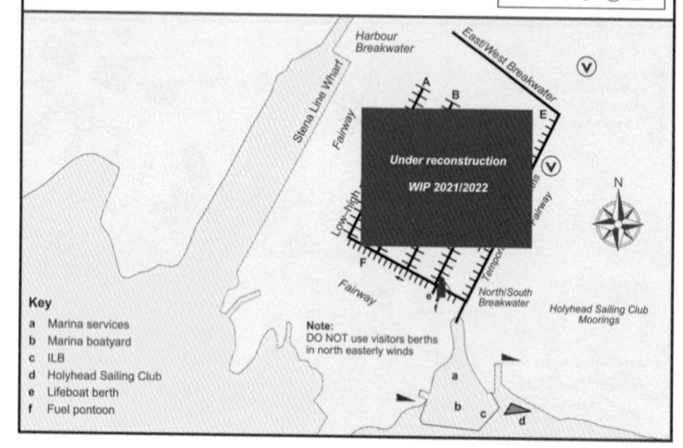
Harbour Breakwater
East/West Breakwater
Stena Line Wharf
Fairway
Under reconstruction
WIP 2021/2022
Low-high
North/South Breakwater
Holyhead Sailing Club Moorings
N
Key
a Marina services
b Marina boatyard
c ILB
d Holyhead Sailing Club
e Lifeboat berth
f Fuel pontoon
Note:
DO NOT use visitors berths in north easterly winds

PWLLHELI MARINA

Pwllheli Marina
Glan Don, Pwllheli, North Wales, LL53 5YT
Tel: 01758 701219
Email: hafanpwllheli@gwynedd.llwy.cymru

VHF	Ch 80
ACCESS	

Pwllheli is an old Welsh market town providing the gateway to the Llyn Peninsula, which stretches out as far as Bardsey Island to form an 'Area of Outstanding Natural Beauty'. Enjoying the spectacular backdrop of the Snowdonia Mountains, Pwllheli's numerous attractions include an open-air market every Wednesday, 'Neuadd Dwyfor', offering a mix of live theatre and latest films, and beautiful beaches.

Pwllheli Marina is situated on the south side of the Llyn Peninsula. One of Wales' finest marinas and sailing centres, it has over 400 pontoon berths and excellent onshore facilities.

FACILITIES AT A GLANCE

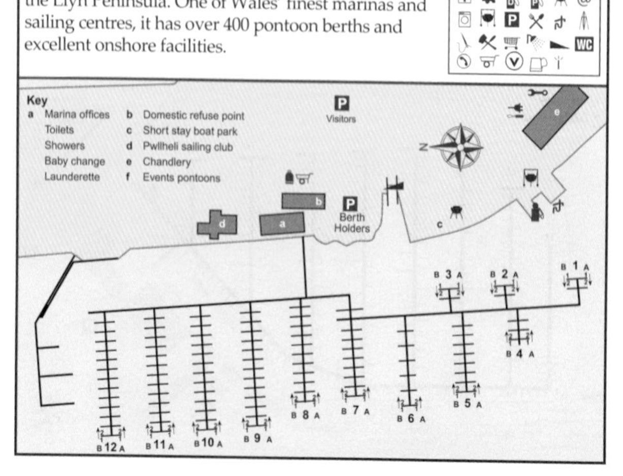
Key
a Marina offices
Toilets
Showers
Baby change
Launderette
b Domestic refuse point
c Short stay boat park
d Pwllheli sailing club
e Chandlery
f Events pontoons
P Visitors
P Berth Holders
B 1 A
B 2 A
B 3 A
B 4 A
B 5 A
B 6 A
B 7 A
B 8 A
B 9 A
B 10 A
B 11 A
B 12 A

SOUTH WALES & BRISTOL CHANNEL – Bardsey Island to Land's End

Key to Marina Plans symbols

- Bottled gas
- Chandler
- Disabled facilities
- Electrical supply
- Electrical repairs
- Engine repairs
- First Aid
- Fresh Water
- Fuel - Diesel
- Fuel - Petrol
- Hardstanding/boatyard
- Internet Café
- Laundry facilities
- Lift-out facilities
- Parking
- Pub/Restaurant
- Pump out
- Rigging service
- Sail repairs
- Shipwright
- Shop/Supermarket
- Showers
- Slipway
- Toilets
- Telephone
- Trolleys
- Visitors berths
- Wi-Fi

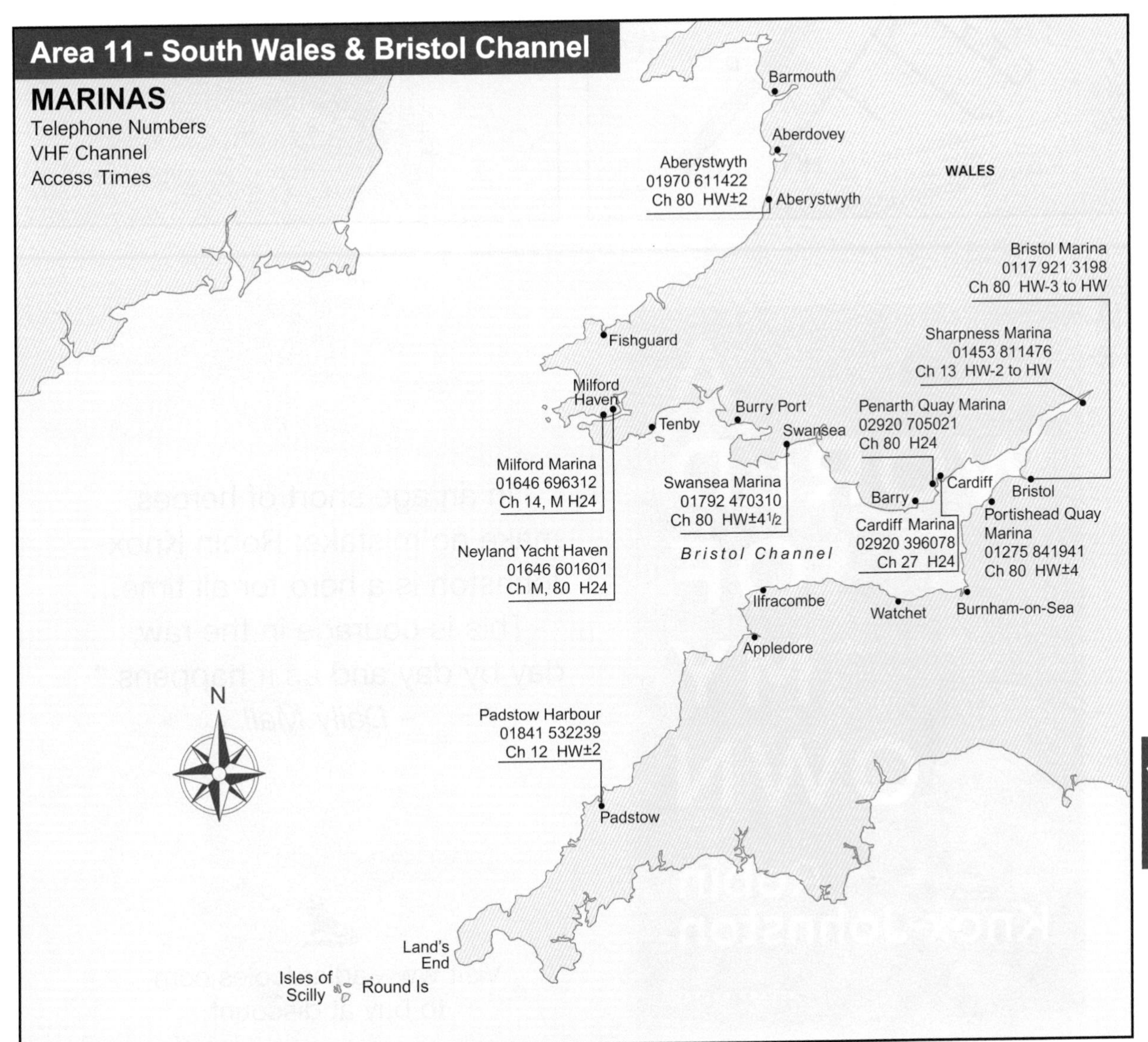
Area 11 - South Wales & Bristol Channel
MARINAS
Telephone Numbers
VHF Channel
Access Times
Barmouth
Aberdovey
Aberystwyth 01970 611422 Ch 80 HW±2
Aberystwyth
WALES
Bristol Marina 0117 921 3198 Ch 80 HW-3 to HW
Fishguard
Sharpness Marina 01453 811476 Ch 13 HW-2 to HW
Milford Haven
Burry Port
Penarth Quay Marina 02920 705021 Ch 80 H24
Tenby
Swansea
Milford Marina 01646 696312 Ch 14, M H24
Swansea Marina 01792 470310 Ch 80 HW±4½
Cardiff
Bristol
Barry
Portishead Quay Marina 01275 841941 Ch 80 HW±4
Cardiff Marina 02920 396078 Ch 27 H24
Neyland Yacht Haven 01646 601601 Ch M, 80 H24
Bristol Channel
Ilfracombe
Watchet
Burnham-on-Sea
Appledore
N
Padstow Harbour 01841 532239 Ch 12 HW±2
Padstow
Land's End
Isles of Scilly
Round Is

ABERYSTWYTH MARINA

Aberystwyth Marina
Trefechan, Aberystwyth, Ceredigion, SY23 1AS
Tel: 01970 611422 Fax: 01970 624122
Email: aber@themarinegroup.co.uk

VHF	Ch 80
ACCESS	HW±2

Aberystwyth Marina offers 165 first class berths providing safe, secure and sheltered moorings for motor boats and yachts.

The onsite chandlery has a selection of clothing, safety equipment, ropes, maintenance, navigation and electrical equipment. The brokerage has a range of motor boats and yachts for sale. In addition, the marina offers a range of boatyard and engine servicing. Other facilities include a slipway and 10t hoist.

Aberystwyth has a range of cafes, seaside fish and chip shops, restaurants, pubs and bars, many a short walk from the marina. The seafront, Promenade and pier is a great location for a walk and to look out over the Irish Sea.

FACILITIES AT A GLANCE

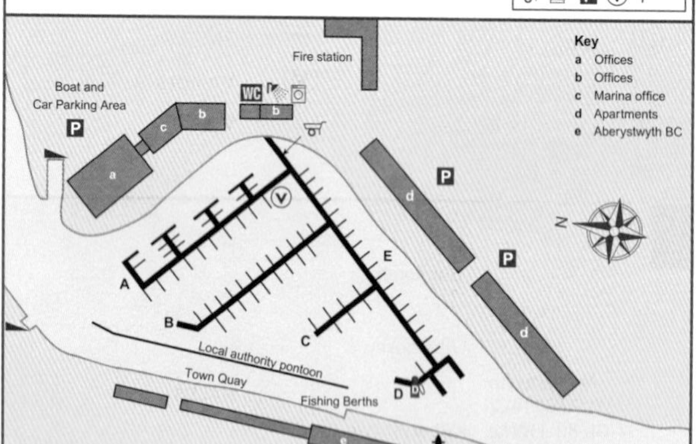
Fire station
Boat and
Car Parking Area
Key
a Offices
b Offices
c Marina office
d Apartments
e Aberystwyth BC
A
B
C
D
E
N
Local authority pontoon
Town Quay
Fishing Berths

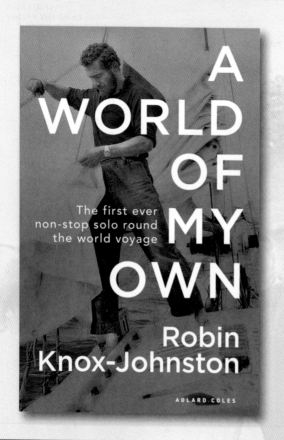

MILFORD MARINA

Milford Marina
Milford Docks, Milford Haven, Pembrokeshire, SA73 3AF
Tel: 01646 696312 Fax: 01646 696314
Email: enquiries@milfordmarina.com www.milfordmarina.com

VHF	Ch 14
ACCESS	H24

Set within one of the deepest natural harbours in the world, Milford Marina is situated in a non-tidal basin within the UK's only coastal National Park, the ideal base for discovering the fabulous coastline in Pembrokeshire, Wales and Ireland. Continued investment has meant that the marina provides safe, secure and sheltered boat berths with a full range of shoreside facilities in the heart of SW Wales.

Accessed via an entrance lock (with waiting pontoons both inside and outside the lock), the marina is perfect for exploring the picturesque upper reaches of the River Cleddau or cruising out beyond St Ann's Head to the unspoilt islands of Skomer, Skokholm and Grassholm.

FACILITIES AT A GLANCE

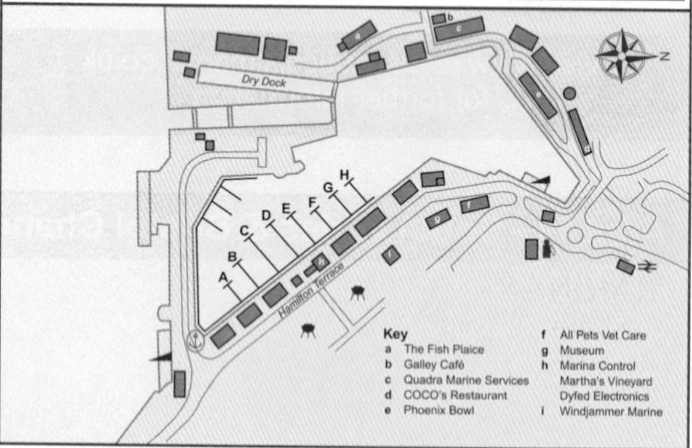
Dry Dock
Hamilton Terrace
A
B
C
D
E
F
G
H
N
Key
a The Fish Plaice
b Galley Café
c Quadra Marine Services
d COCO's Restaurant
e Phoenix Bowl
f All Pets Vet Care
g Museum
h Marina Control
Martha's Vineyard
Dyfed Electronics
i Windjammer Marine

NEYLAND YACHT HAVEN

Neyland Yacht Haven Ltd
Brunel Quay, Neyland, Pembrokeshire, SA73 1PY
Tel: 01646 601601
Email: neyland@yachthavens.com www.yachthavens.com

VHF	Ch M, 80
ACCESS	H24

Approximately 10 miles from the entrance to Milford Haven lies Neyland Yacht Haven. Tucked away in a well protected inlet just before the Cleddau Bridge, this marina has 420 berths and can accommodate yachts up to 25m LOA with draughts of up to 2.5m. The marina is divided into two basins, with the lower one enjoying full tidal access, while entry to the upper one is restricted by a tidal sill. Visitor and annual berthing enquiries welcome.

Offering a comprehensive range of services, Neyland Yacht Haven is within a five minute walk of the town centre where the various shops and takeaways cater for most everyday needs; bicycle hire is also available. The Yacht Haven is a member of the TransEurope Group.

FACILITIES AT A GLANCE

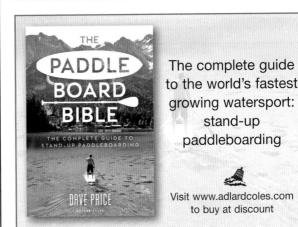

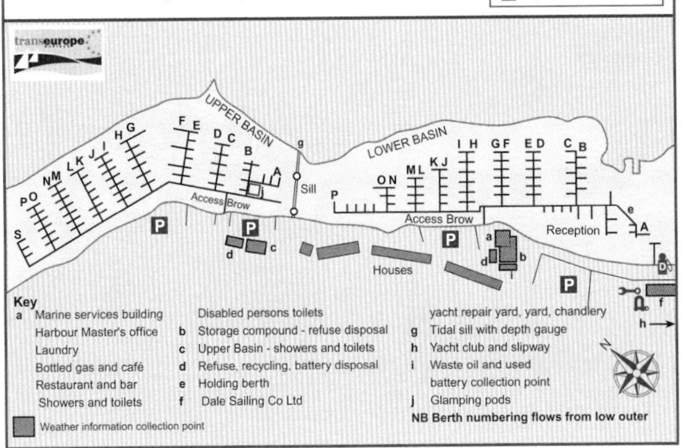
transeurope
UPPER BASIN
LOWER BASIN
Sill
Access Brow
Access Brow
Reception
Houses
Key
a Marine services building
Harbour Master's office
Laundry
Bottled gas and café
Restaurant and bar
Showers and toilets
Disabled persons toilets
b Storage compound - refuse disposal
c Upper Basin - showers and toilets
d Refuse, recycling, battery disposal
e Holding berth
f Dale Sailing Co Ltd
yacht repair yard, yard, chandlery
g Tidal sill with depth gauge
h Yacht club and slipway
i Waste oil and used battery collection point
j Glamping pods
NB Berth numbering flows from low outer
Weather information collection point

SWANSEA MARINA

Swansea Marina
Lockside, Maritime Quarter, Swansea, SA1 1WG
Tel: 01792 470310 Fax: 01792 463948
www.swansea.gov.uk/swanseamarina
Email: swanmar@swansea.gov.uk

VHF	Ch 80
ACCESS	HW±4.5

At the hub of the city's redeveloped and award winning Maritime Quarter, Swansea Marina can be accessed HW±4½ hrs via a lock. Surrounded by a plethora of shops, restaurants and marine businesses to cater for most yachtsmen's needs, the marina is in close proximity to the picturesque Gower coast, where there is no shortage of quiet sandy beaches off which to anchor. It also provides the perfect starting point for cruising to Ilfracombe, Lundy Island, the North Cornish coast or West Wales.

Within easy walking distance of the marina is the city centre, boasting a covered shopping centre and market. For those who prefer walking or cycling, take the long promenade to the Mumbles fishing village from where there are plenty of coastal walks.

FACILITIES AT A GLANCE

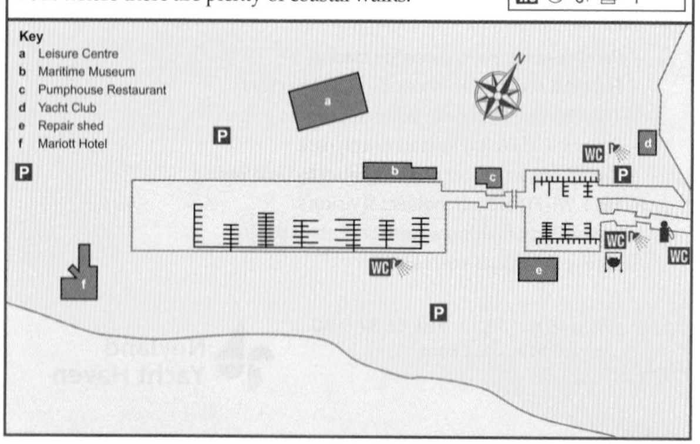
Key
a Leisure Centre
b Maritime Museum
c Pumphouse Restaurant
d Yacht Club
e Repair shed
f Mariott Hotel
N
P
WC

PENARTH MARINA

Penarth Marina
Penarth, Vale of Glamorgan, CF64 1TQ
Tel: 02920 705021
Email: penarth@boatfolk.co.uk
www.boatfolk.co.uk/penarthmarina

VHF	Ch 80
ACCESS	H24

Penarth Marina has been established in the historic basins of Penarth Docks for over 20 years and is the premier boating facility in the region. The marina is Cardiff Bay's only 5 Gold Anchor marina and provides an ideal base for those using the Bay and the Bristol Channel.

With 24hr access there is always water available for boating. Penarth and Cardiff boast an extensive range of leisure facilities, shops and restaurants making this marina an ideal base or destination. The marina has a blue flag and is a member of the TransEurope Group.

FACILITIES AT A GLANCE

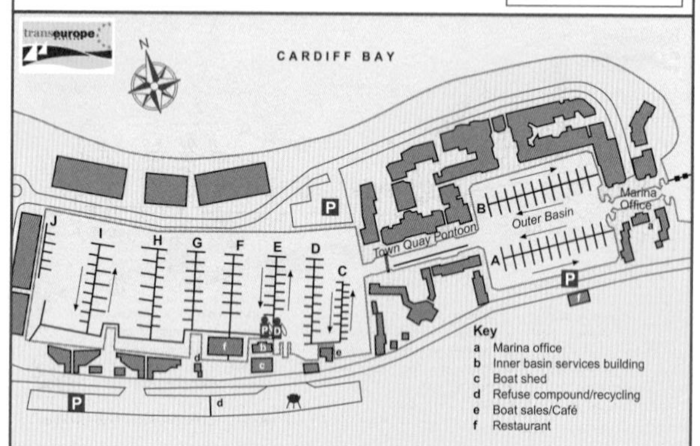
transeurope
N
CARDIFF BAY
Town Quay Pontoon
Outer Basin
Marina Office
J
I
H
G
F
E
D
C
B
A
P
Key
a Marina office
b Inner basin services building
c Boat shed
d Refuse compound/recycling
e Boat sales/Café
f Restaurant

CARDIFF MARINA

Cardiff Marina
Watkiss Way, Cardiff, CF11 0SY
Tel: 02920 396078 Fax: 02920 345116
Email: info@themarinegroup.co.uk
www.themarinegroup.co.uk

VHF	Ch M
ACCESS	H24

A sheltered haven on the River Ely, Cardiff Marina offers 350 fully serviced berths within Cardiff Bay. Adjacent to Cardiff Marina, the waterside setting of Bayscape, a new mixed-use development of 115 apartments, is also home to the brand-new marina management suite, laundry and washrooms. The south facing terrace of the lounge bar, is a great location for berth holders to relax and unwind. Marina facilities also include a 50t Sealift, platform crane for mast work and brokerage. In addition, Cardiff Marine Village is home to Cardiff Marine Services, Wales' leading boatyard and refit and repair centre. The 3-acre site also has ample space for hard standing storage.

FACILITIES AT A GLANCE

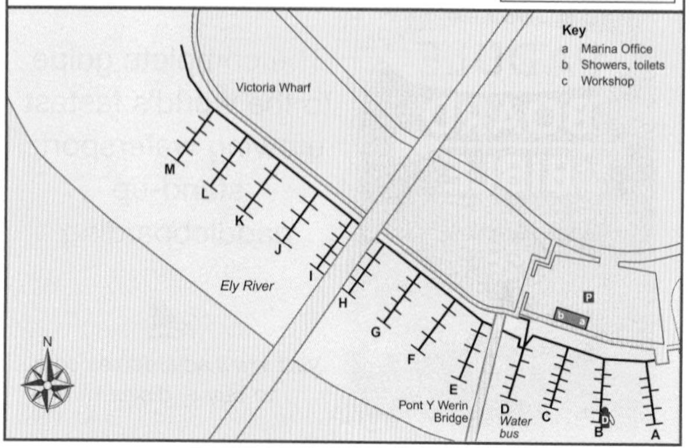
Key
a Marina Office
b Showers, toilets
c Workshop
Victoria Wharf
Ely River
M
L
K
J
I
H
G
F
E
D
C
B
A
Pont Y Werin Bridge
Water bus
N
P

BRISTOL MARINA

Bristol Marina Ltd
Hanover Place, Bristol, BS1 6UH
Tel: 0117 921 3198 Fax: 0117 929 7672
Email: info@bristolmarina.co.uk

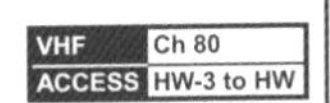
VHF Ch 80
ACCESS HW-3 to HW

Situated in the heart of the city, Bristol is a fully serviced marina providing over 100 pontoon berths for vessels up to 20m LOA. Among the facilities are a new fuelling berth and pump out station as well as an on site chandler and sailmaker. It is situated on the south side of the Floating Harbour, about eight miles from the mouth of the River Avon. Accessible from seaward via the Cumberland Basin, passing through both Entrance Lock and Junction Lock, it can be reached approximately three hours before HW.

Shops, restaurants, theatres and cinemas are all within easy reach of the marina, while local attractions include the SS *Great Britain*, designed by Isambard Kingdom Brunel, and the famous Clifton Suspension Bridge, which has an excellent visitors' centre depicting its fascinating story.

FACILITIES AT A GLANCE

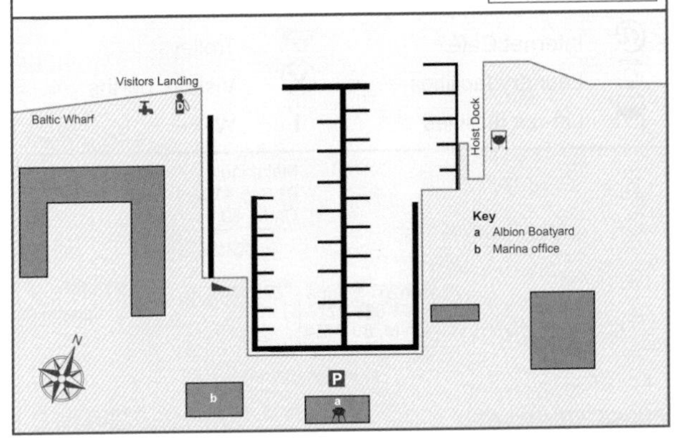
Visitors Landing
Baltic Wharf
Hoist Dock
Key
a Albion Boatyard
b Marina office
N

PORTISHEAD MARINA

Portishead Marina
Newfoundland Way, Portishead, North Somerset, BS20 7DF
Tel: 01275 841941
Email: portishead@boatfolk.co.uk
www.boatfolk.co.uk/portisheadmarina

VHF Ch 80
ACCESS HW±3.5

Portishead Marina is a popular destination for cruising in the Bristol Channel. Providing an excellent link between the inland waterways at Bristol and Sharpness and offering access to the open water and marinas down channel. The entrance to the marina is via a lock, with a minimum access of HW+/- 3.5hrs. Contact the marina on Ch 80 ahead of time for next available lock. The marina provides 320 fully serviced berths and can accommodate vessels up to 40m LOA, draft up to 5.5m. The marina has a 35 tonnes boat hoist and the boatyard offers all the facilities you would expect from a boatfolk site.

FACILITIES AT A GLANCE

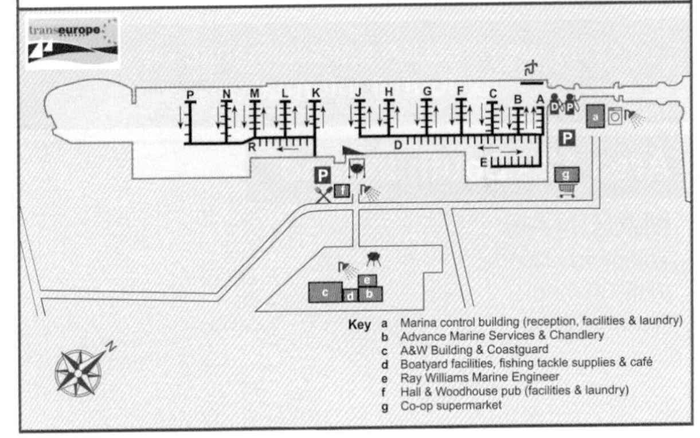
transeurope
P N M L K J H G F C B A
R D E
Key a Marina control building (reception, facilities & laundry)
b Advance Marine Services & Chandlery
c A&W Building & Coastguard
d Boatyard facilities, fishing tackle supplies & café
e Ray Williams Marine Engineer
f Hall & Woodhouse pub (facilities & laundry)
g Co-op supermarket
N

PADSTOW HARBOUR

Padstow Harbour Commissioners
The Harbour Office, Padstow, Cornwall, PL28 8AQ
Tel: 01841 532239 Fax: 01841 533346
Email: harbourmaster@padstow-harbour.co.uk
www.padstow-harbour.co.uk

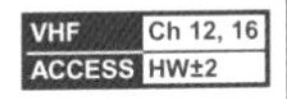
VHF Ch 12, 16
ACCESS HW±2

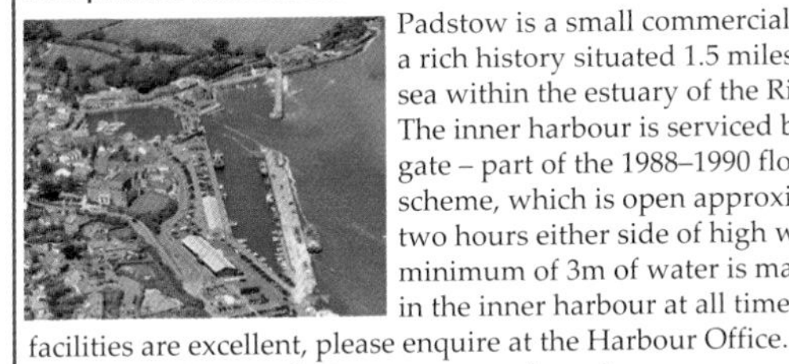

Padstow is a small commercial port with a rich history situated 1.5 miles from the sea within the estuary of the River Camel. The inner harbour is serviced by a tidal gate – part of the 1988–1990 flood defence scheme, which is open approximately two hours either side of high water. A minimum of 3m of water is maintained in the inner harbour at all times. Onshore facilities are excellent, please enquire at the Harbour Office.

This is a thriving fishing port, so perhaps it wasn't surprising that celebrity chefs like Rick Stein and Paul Ainsworth would set up shop in the town. This is a great resting point for everything Cornish from surf and coastal walks to fish and chips and cream teas.

FACILITIES AT A GLANCE

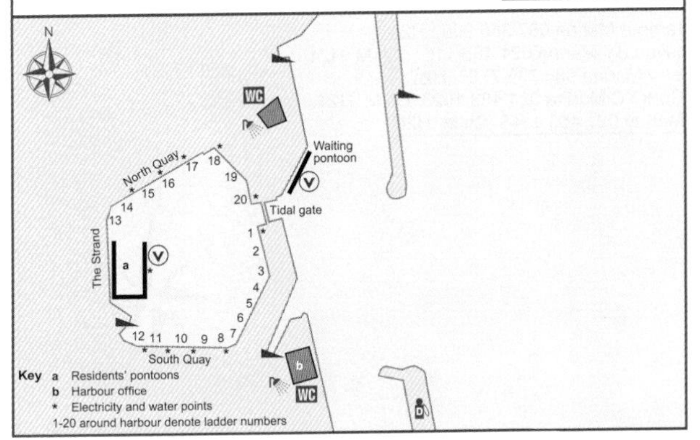
N
WC
North Quay
Waiting pontoon
Tidal gate
The Strand
South Quay
Key a Residents' pontoons
b Harbour office
* Electricity and water points
1-20 around harbour denote ladder numbers

SOUTH IRELAND – Malahide, clockwise to Liscannor Bay

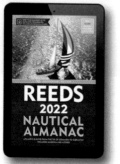

Key to Marina Plans symbols

- Bottled gas
- Chandler
- Disabled facilities
- Electrical supply
- Electrical repairs
- Engine repairs
- First Aid
- Fresh Water
- Fuel - Diesel
- Fuel - Petrol
- Hardstanding/boatyard
- Internet Café
- Laundry facilities
- Lift-out facilities
- Parking
- Pub/Restaurant
- Pump out
- Rigging service
- Sail repairs
- Shipwright
- Shop/Supermarket
- Showers
- Slipway
- Toilets
- Telephone
- Trolleys
- Visitors berths
- Wi-Fi

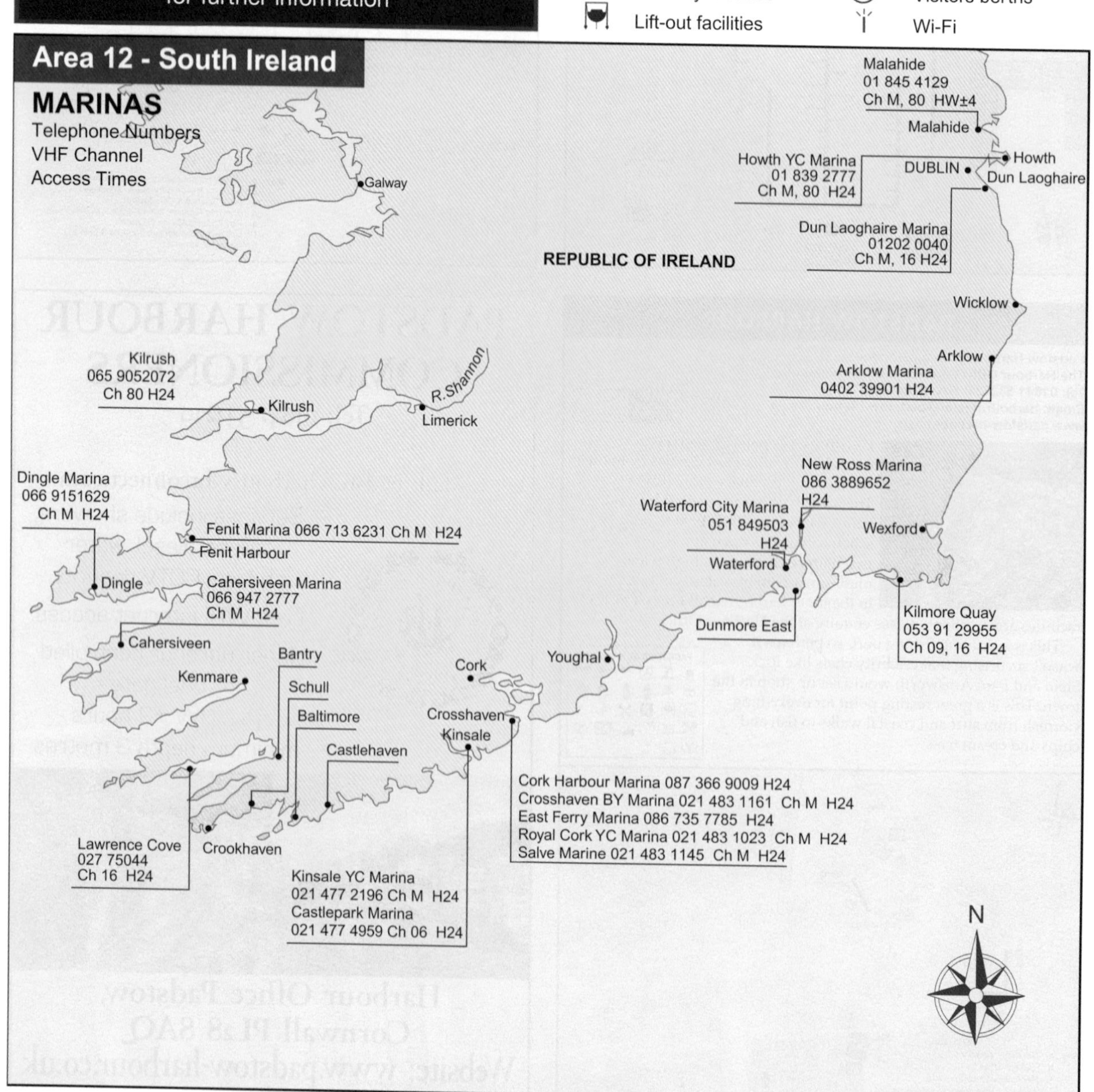
Area 12 - South Ireland
MARINAS
Telephone Numbers
VHF Channel
Access Times
Galway
REPUBLIC OF IRELAND
Malahide
01 845 4129
Ch M, 80 HW±4
Malahide
Howth YC Marina
01 839 2777
Ch M, 80 H24
DUBLIN
Howth
Dun Laoghaire
Dun Laoghaire Marina
01202 0040
Ch M, 16 H24
Wicklow
Arklow
Arklow Marina
0402 39901 H24
Kilrush
065 9052072
Ch 80 H24
Kilrush
R.Shannon
Limerick
New Ross Marina
086 3889652
H24
Waterford City Marina
051 849503
H24
Waterford
Wexford
Dingle Marina
066 9151629
Ch M H24
Fenit Marina 066 713 6231 Ch M H24
Fenit Harbour
Dingle
Cahersiveen Marina
066 947 2777
Ch M H24
Cahersiveen
Dunmore East
Kilmore Quay
053 91 29955
Ch 09, 16 H24
Bantry
Kenmare
Schull
Baltimore
Castlehaven
Cork
Youghal
Crosshaven
Kinsale
Cork Harbour Marina 087 366 9009 H24
Crosshaven BY Marina 021 483 1161 Ch M H24
East Ferry Marina 086 735 7785 H24
Royal Cork YC Marina 021 483 1023 Ch M H24
Salve Marine 021 483 1145 Ch M H24
Lawrence Cove
027 75044
Ch 16 H24
Crookhaven
Kinsale YC Marina
021 477 2196 Ch M H24
Castlepark Marina
021 477 4959 Ch 06 H24
N

MALAHIDE MARINA

Malahide Marina
Malahide, Co. Dublin
Tel: +353 1 845 4129 Fax: +353 1 845 4255
Email: info@malahidemarina.net
www.malahidemarina.net

VHF	Ch M, 80
ACCESS	HW±4

Malahide Marina, situated just 10 minutes from Dublin Airport and 20 minutes north of Dublin's city centre, is a fully serviced marina accommodating up to 350 yachts. Capable of taking vessels of up to 75m in length, its first class facilities include a boatyard with hard standing for approximately 170 boats and a 30-ton mobile hoist. Its on site restaurant provides a large seating area in convivial surroundings. The village of Malahide has plenty to offer the visiting yachtsmen, with a wide variety of eating places, nearby golf courses and tennis courts as well as a historic castle and botanical gardens.

FACILITIES AT A GLANCE

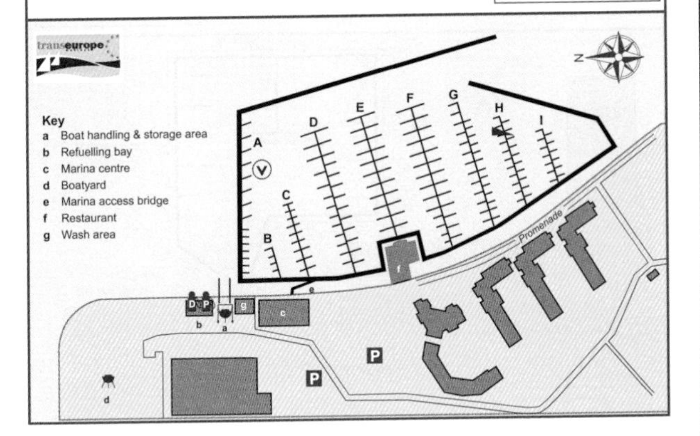
transeurope
Key
a Boat handling & storage area
b Refuelling bay
c Marina centre
d Boatyard
e Marina access bridge
f Restaurant
g Wash area
A
B
C
D
E
F
G
H
I
Promenade
P

HOWTH MARINA

Howth Marina
Howth Marina, Harbour Road, Howth, Co. Dublin
Tel: +353 1 8392777 Fax: +353 1 8392430
Email: marina@hyc.ie
www.hyc.ie

VHF	Ch M, 80
ACCESS	H24

Based on the north coast of the rugged peninsula that forms the northern side of Dublin Bay, Howth Marina is ideally situated for north or south-bound traffic in the Irish Sea. Well sheltered in all winds, it can be entered at any state of the tide. Overlooking the marina is Howth Yacht Club, which has in recent years been expanded and is now said to be the largest yacht club in Ireland. With good road and rail links, Howth is in easy reach of Dublin's airport and ferry terminal, making it an obvious choice for crew changeovers.

FACILITIES AT A GLANCE

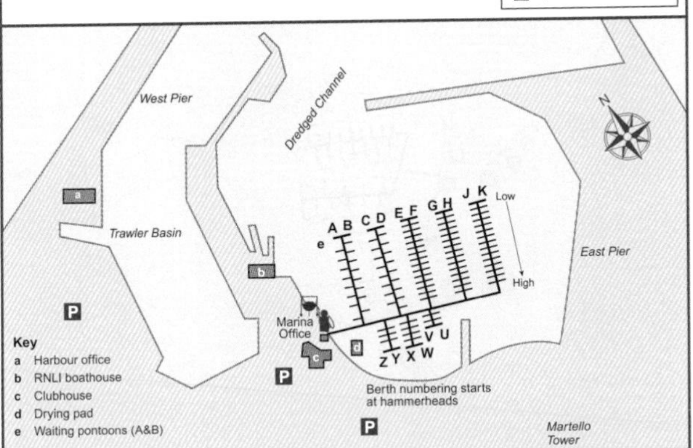
West Pier
Dredged Channel
Trawler Basin
East Pier
A B C D E F G H J K
Low
High
e
Marina Office
Z Y X W V U
Berth numbering starts at hammerheads
Martello Tower
Key
a Harbour office
b RNLI boathouse
c Clubhouse
d Drying pad
e Waiting pontoons (A&B)

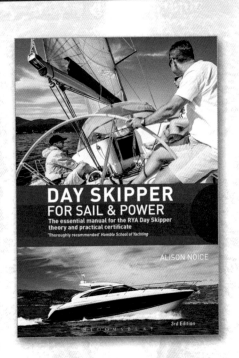

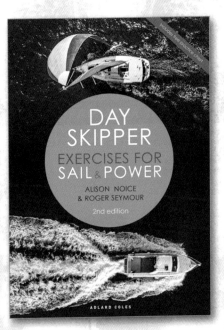

DUN LAOGHAIRE MARINA

Dun Laoghaire Marina
Harbour Road, Dun Laoghaire, Co Dublin, Ireland
Tel: +353 1 202 0040
Email: info@dlmarina.com www.dlmarina.com

VHF	Ch M, 16
ACCESS	H24

Dun Laoghaire Marina – Gateway to Dublin and the first marina in the Republic of Ireland to be awarded five Gold anchors by THYA and also achieved the ICOMIA 'Clean Marina' accreditation – is the largest marina in Ireland. 24 hour access in all weather conditions. The town centre is located within 400m. Serviced berthing for 820 boats from 6m to 23m with visitors mainly on the hammerheads.

Larger vessels up to 46m and 160 tonnes can be berthed alongside breakwater pontoons. Minimum draft is 3.8m LWS.

Three phase power is available. With Dublin rail station 12km and airport 35km, this is an ideal base for Irish culture and entertainment, as well as crew changes etc. Easy access to Dublin Bay, home to the biennial Dun Laoghaire Regatta.

FACILITIES AT A GLANCE
WC

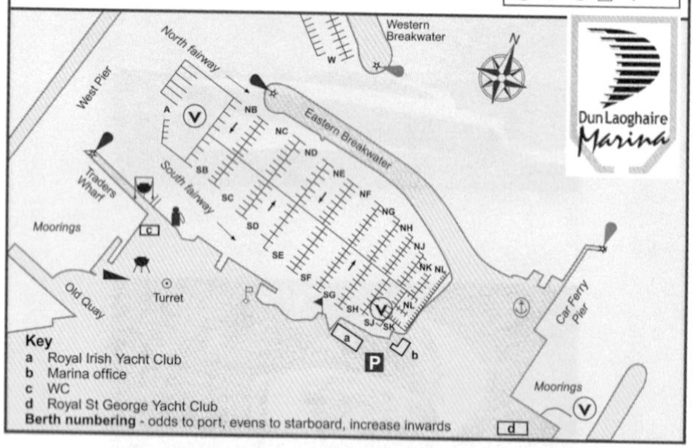
Western Breakwater
North fairway
West Pier
Eastern Breakwater
South fairway
Traders Wharf
Moorings
Old Quay
Turret
Car Ferry Pier
Moorings
Dun Laoghaire Marina
Key
a Royal Irish Yacht Club
b Marina office
c WC
d Royal St George Yacht Club
Berth numbering - odds to port, evens to starboard, increase inwards

ARKLOW MARINA

Arklow Marina
North Quay, Arklow, Co. Wicklow, Ireland
Mobiles: 087 2375189 or 087 2588078
Email: personnel@asl.ie
www.arklowmarina.com

VHF	
ACCESS	H24

Arklow is a popular fishing port and seaside town situated at the mouth of the River Avoca, 16 miles south of Wicklow and 11 miles north east of Gorey. The town is ideally placed for visiting the many beauty spots of County Wicklow including Glenmalure, Glendalough and Clara Lara, Avoca (Ballykissangel).

Arklow Marina is on the north bank of the river (dredged 2014) just upstream of the commercial quays, with 42 berths in an inner harbour and 30 berths on pontoons outside the marina entrance. Vessels over 14m LOA should moor on the river pontoons.

FACILITIES AT A GLANCE

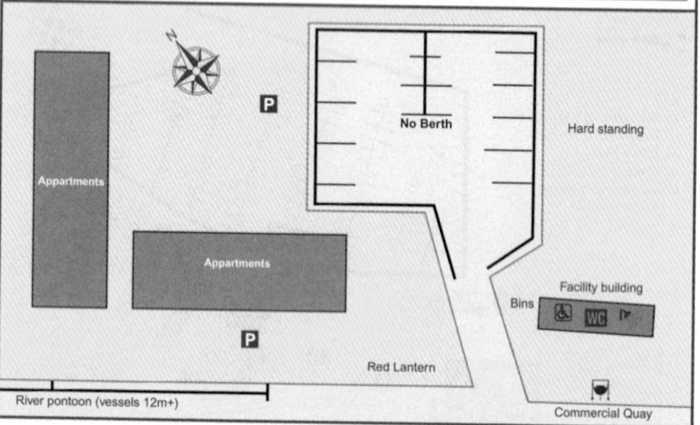
No Berth
Hard standing
Appartments
Appartments
Facility building
Bins
WC
Red Lantern
River pontoon (vessels 12m+)
Commercial Quay

NEW ROSS MARINA

New Ross Marina
The Tholsel, Quay Street, New Ross, Co Wexford, Y34 CF64
Tel: +353 86 3889652
www.newrossmarina.com
Email: newrossmarina@wexfordcoco.ie

VHF	
ACCESS	H24

Situated on the River Barrow, New Ross Marina is an ideal base for those sailing the south and east of Ireland. A 60-berth marina located 18M from Hook Head, New Ross is a short hop from Kilmore Quay, Dunmore East and Waterford City Marinas.

The River Barrow is the second longest river in Ireland and an important link to the Inland Waterways Network throughout the island, connecting New Ross to Dublin, Limerick, Carrick on Shannon, Enniskillen and the NW. The replica famine ship Dunbrody is a stroll up the quays, and the Kennedy Homestead just a few miles down the road.

FACILITIES AT A GLANCE
WC

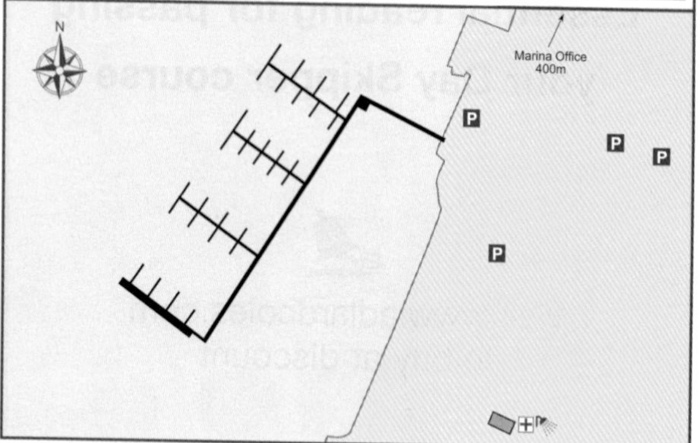
N
Marina Office 400m

KILMORE QUAY

Kilmore Quay
Wexford, Ireland
Tel: +353 53 9129955 www.kilmorequaymarina.com
Email: assistant.marineofficer@wexfordcoco.ie

VHF	Ch 09, 16
ACCESS	H24

Located in the SE corner of Ireland, Kilmore Quay is a small rural fishing village situated approximately 14 miles from the town of Wexford and 12 miles from Rosslare ferry port.

Its 55-berthed marina, offering shelter from the elements as well as various on shore facilities, including diesel available 24/7, has become a regular port of call for many cruising yachtsmen. With several nearby areas of either historical or natural significance accessible using local bike hire, Kilmore is renowned for its 'green' approach to the environment.

FACILITIES AT A GLANCE

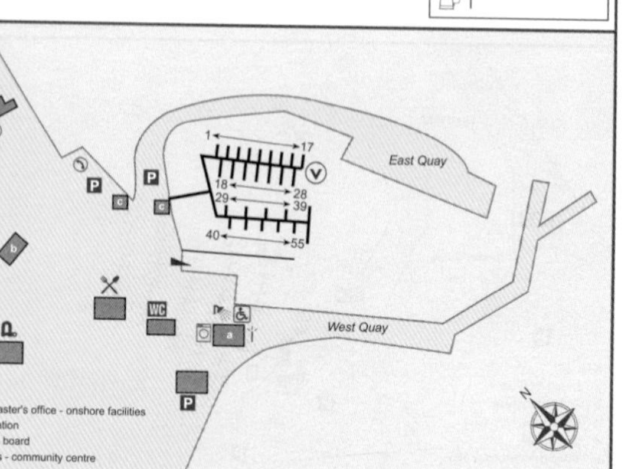
East Quay
West Quay
Key
a Harbour master's office - onshore facilities
b Lifeboat station
c Information board
d Stella Maris - community centre

WATERFORD CITY MARINA

Waterford City Marina
Waterford, Ireland
Tel: +353 87 238 4944
Email: jcodd@waterfordcouncil.ie

VHF
ACCESS H24

Famous for its connections with Waterford Crystal, manufactured in the city centre, Waterford is the capital of the SE region of Ireland. As a major city, it benefits from good rail links with Dublin, and Limerick, a regional airport with daily flights to Britain and an extensive bus service to surrounding towns and villages. The marina is found on the banks of the River Suir, in the heart of this historic Viking city dating back to the ninth century. Yachtsmen can make the most of Waterford's wide range of shops, restaurants and bars without having to walk too far from their boats. With 100 fully serviced berths and first rate security, Waterford City Marina now provides shower, toilet and laundry facilities in its new reception building.

FACILITIES AT A GLANCE
WC

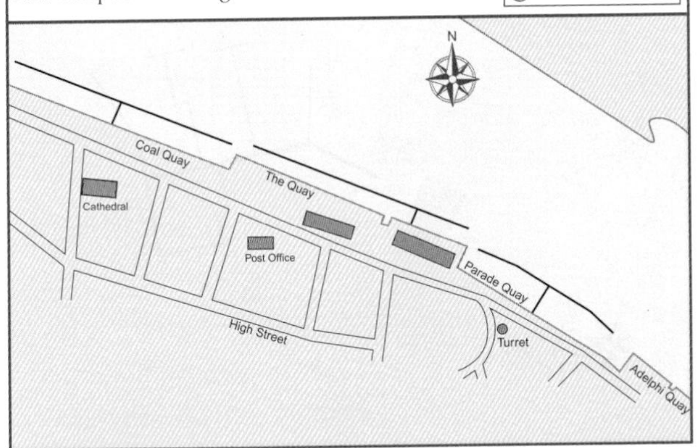
N
Coal Quay
The Quay
Cathedral
Post Office
Parade Quay
High Street
Turret
Adelphi Quay

CROSSHAVEN BOATYARD MARINA

Crosshaven Boatyard Marina
Crosshaven, Co Cork, Ireland
Tel: +353 214 831161 Fax: +353 214 831603
Email: info@crosshavenboatyard.com

VHF Ch M
ACCESS H24

One of three marinas at Crosshaven, Crosshaven Boatyard was founded in 1950 and originally made its name from the construction of some of the most world-renowned yachts, including *Gypsy Moth* and Denis Doyle's *Moonduster*. Nowadays, however, the yard has diversified to provide a wide range of services to both the marine leisure and professional industries. Situated on a safe and sheltered river only 12 miles from Cork City Centre, the marina boasts 100 fully-serviced berths along with the capacity to accommodate yachts up to 35m LOA with a 4m draught. In addition, it is ideally situated for cruising the stunning south west coast of Ireland.

FACILITIES AT A GLANCE
WC

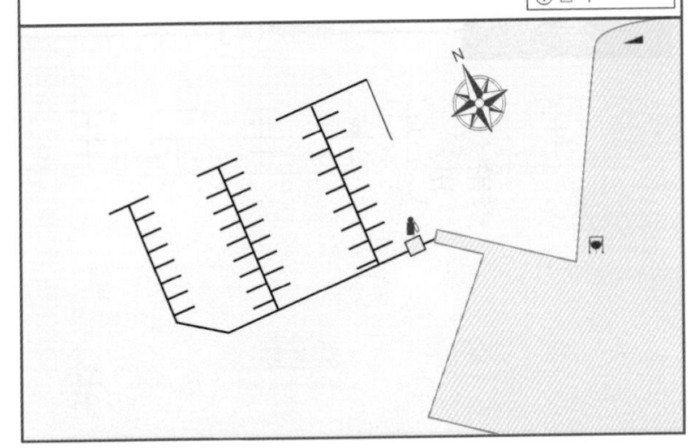
N

SALVE MARINE

Salve Marine
Crosshaven, Co Cork, Ireland
Tel: +353 21 483 1145 Fax: +353 21 483 1747
Email: salvemarineltd@gmail.com

VHF Ch M
ACCESS H24

Crosshaven is a picturesque seaside resort providing a gateway to Ireland's south and south west coasts. Offering a variety of activities to suit all types, its rocky coves and quiet sandy beaches stretch from Graball to Church Bay and from Fennell's Bay to nearby Myrtleville. Besides a selection of craft shops selling locally produced arts and crafts, there are plenty of pubs, restaurants and takeaways to suit even the most discerning of tastes. Lying within a few hundred metres of the village centre is Salve Marine, accommodating yachts up to 43m LOA with draughts of up to 4m. Its comprehensive services range from engineering and welding facilities to hull and rigging repairs.

FACILITIES AT A GLANCE
WC

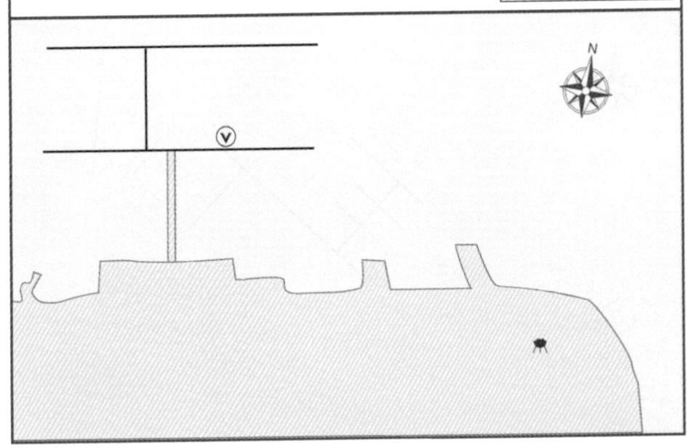
N

CORK HARBOUR MARINA

Cork Harbour Marina
Monkstown, Co Cork, Ireland
Tel: +353 87 366 9009
Email: info@corkharbourmarina.ie www.corkharbourmarina.ie

VHF
ACCESS H24

Located in the picturesque town of Monkstown, Cork Harbour Marina is in the heart of Cork Harbour. The marina can cater for all boat types with a draft of up to 7m and the marina is accessible at all phases of the tide.

Within strolling distance from the marina is the 'Bosun' restaurant and 'Napoli', an Italian delicatessen. There is also a sailing club, tennis club and golf club located nearby. Monkstown is just a short riverside walk from Passage West, with all the amenities one might require.

There is a frequent bus service from the marina gates to Cork City. Cork Harbour Marina is 15km from Cork International Airport.

FACILITIES AT A GLANCE
WC

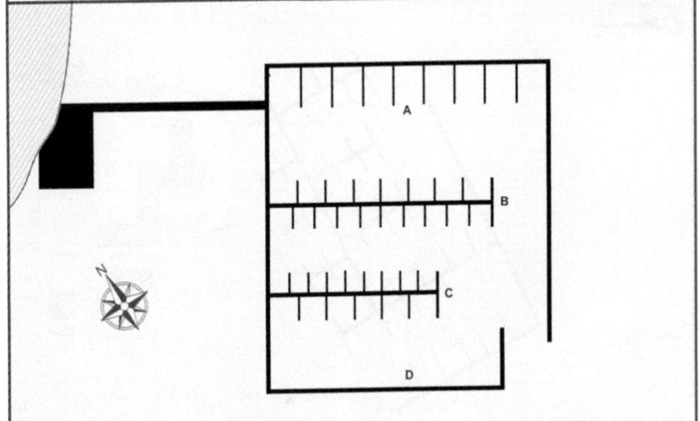
A
B
C
D
N

ROYAL CORK YACHT CLUB

Royal Cork Yacht Club Marina
Crosshaven, Co Cork, Ireland
Tel: +353 21 483 1023 Fax: +353 21 483 2657
Email: mark@royalcork.com www.royalcork.com

VHF	Ch M
ACCESS	H24

Founded in 1720, the Royal Cork Yacht Club is the oldest and one of the most prominent yacht clubs in the world. Organising, among many other events, the prestigious biennial Volvo Cork Week, it boasts a number of World, European and national sailors among its membership.

The Yacht Club's marina is situated at Crosshaven, on the hillside at the mouth of the Owenabue River, just inside the entrance to Cork Harbour. The harbour is popular with yachtsmen as it is accessible and well sheltered in all weather conditions. It also benefits from the Gulf Stream producing a temperate climate practically all year round.

FACILITIES AT A GLANCE

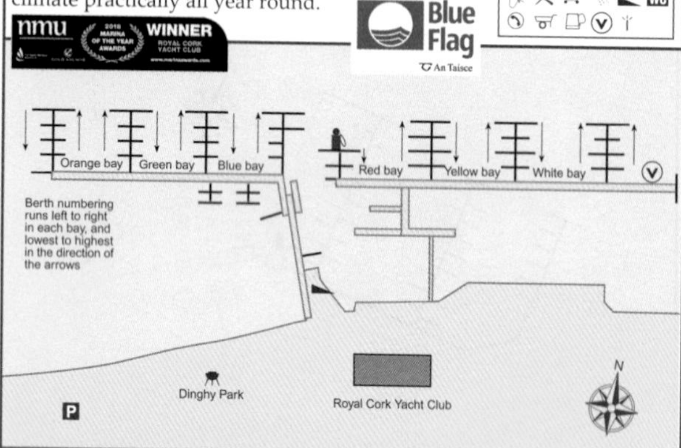
nmu
WINNER
Blue Flag
An Taisce
Orange bay
Green bay
Blue bay
Red bay
Yellow bay
White bay
Berth numbering runs left to right in each bay, and lowest to highest in the direction of the arrows
Dinghy Park
Royal Cork Yacht Club
N

KINSALE YACHT CLUB MARINA

Kinsale Yacht Club Marina
Kinsale, Co Cork, Ireland
Tel: +353 876 787377
Email: kyc@iol.ie

VHF	Ch M
ACCESS	H24

Kinsale is a natural, virtually land-locked harbour on the estuary of the Bandon River, approximately 12 miles south west of Cork harbour entrance. Home to a thriving fishing fleet as well as frequented by commercial shipping, it boasts two fully serviced marinas, with the Kinsale Yacht Club & Marina being the closest to the town. Visitors to this marina automatically become temporary members of the club and are therefore entitled to make full use of the facilities, which include a fully licensed bar and restaurant serving evening meals on Wednesdays, Thursdays and Saturdays. Fuel, water and repair services are also available.

FACILITIES AT A GLANCE

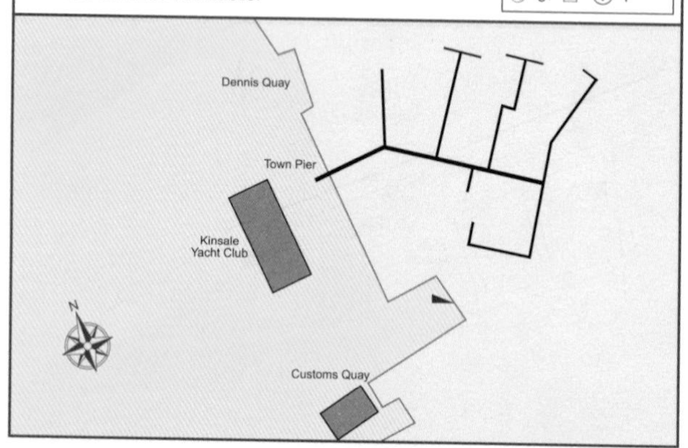
Dennis Quay
Town Pier
Kinsale Yacht Club
Customs Quay
N

CASTLEPARK MARINA

Castlepark Marina Centre
Kinsale, Co Cork, Ireland
Tel: +353 21 4774959
Email: info@castleparkmarina.com

VHF	Ch 16, 14
ACCESS	H24

Situated on the south side of Kinsale Harbour, Castlepark is a small marina with deep water pontoon berths that are accessible at all states of the tide. Surrounded by rolling hills, it boasts its own beach as well as being in close proximity to the parklands of James Fort and a traditional Irish pub. The attractive town of Kinsale, with its narrow streets and slate-clad houses, lies just 1.5 miles away by road or five minutes away by ferry. Known as Ireland's 'fine food centre', it incorporates a number of gourmet food shops and high quality restaurants as well as a wine museum.

FACILITIES AT A GLANCE

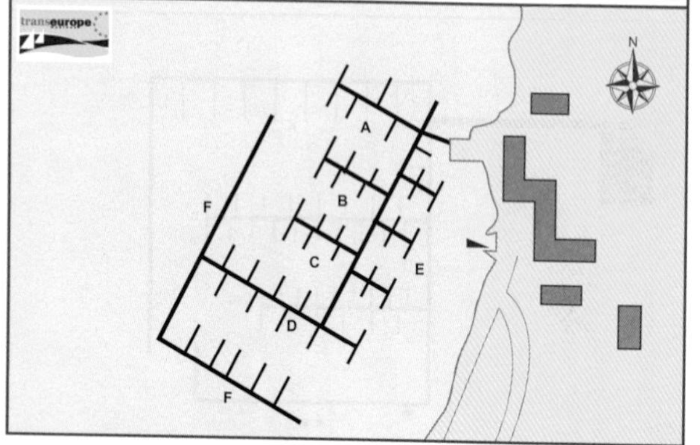
transeurope
A
B
C
D
E
F
F
N

LAWRENCE COVE MARINA

Lawrence Cove Marina
Lawrence Cove, Bere Island, Co Cork, Ireland
Tel: +353 27 75044 mob: +353 879 125930
Email: rachelsherig@gmail.com
www.lawrencecovemarina.ie

VHF	Ch 16
ACCESS	H24

Lawrence Cove enjoys a peaceful location on an island at the entrance to Bantry Bay. Privately owned and run, it offers sheltered and secluded waters as well as excellent facilities and fully serviced pontoon berths. A few hundred yards from the marina you will find a shop, pub and restaurant, while the mainland, with its various attractions, can be easily reached by ferry. Lawrence Cove lies at the heart of the wonderful cruising grounds of Ireland's south west coast and, just two hours from Cork airport, is an ideal place to leave your boat for long or short periods.

FACILITIES AT A GLANCE

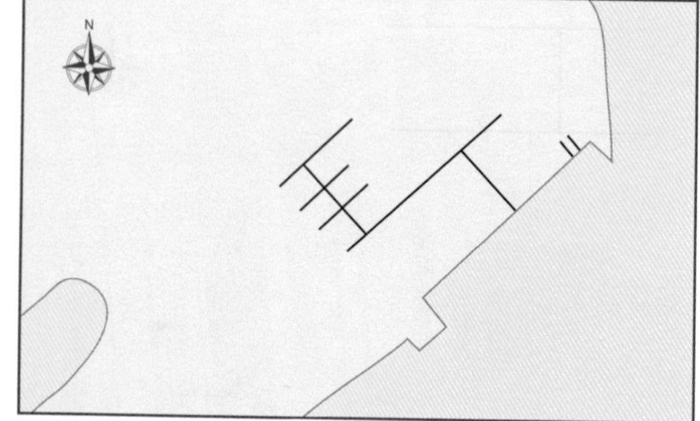
N

CAHERSIVEEN MARINA

Cahersiveen Marina
The Pier, Cahersiveen, Co. Kerry, Ireland
Tel: +353 66 9472777
Email: acardsiveen@gmail.com
www.cahersiveenmarina.ie

VHF	Ch 24
ACCESS	H24

Situated two miles up Valentia River from Valentia Harbour, Cahersiveen Marina is well protected in all wind directions and is convenient for sailing to Valentia Island and Dingle Bay as well as for visiting some of the spectacular uninhabited islands in the surrounding area. Boasting a host of sheltered sandy beaches, the region is renowned for salt and fresh water fishing as well as being good for scuba diving.

Within easy walking distance of the marina lies the historic town of Cahersiveen, incorporating an array of convivial pubs and restaurants.

FACILITIES AT A GLANCE
@
P
WC

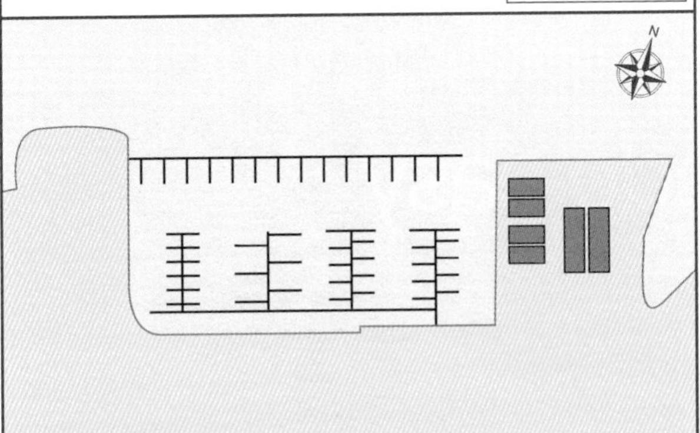
N

DINGLE MARINA

Dingle Marina
c/o Dingle Fishery Harbour Centre, Strand Street, Dingle, Co Kerry, Ireland
Tel: +353 (0)87 925 4115 Fax: +353 (0)69 5152546
Email:dingleharbour@agriculture.gov.ie
www.dinglemarina.ie

VHF	Ch 14
ACCESS	H24

Dingle is Ireland's most westerly marina, lying at the heart of the sheltered Dingle Harbour, and is easily reached both day and night via a well buoyed approach channel. The surrounding area is an interesting and unfrequented cruising ground, with several islands, bays and beaches for the yachtsman to explore.

The marina lies in the heart of the old market town, renowned for its hospitality and traditional Irish pub music. Besides enjoying the excellent seafood restaurants and 52 pubs, other recreational pastimes include horse riding, golf, climbing and diving.

FACILITIES AT A GLANCE
@
WC

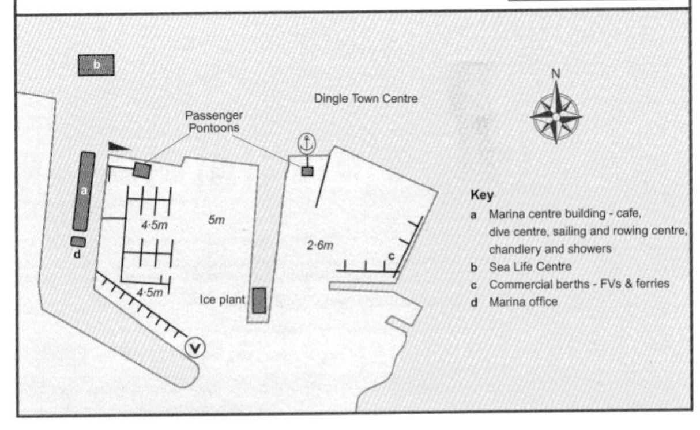
b
Dingle Town Centre
Passenger Pontoons
N
a
d
4·5m
5m
2·6m
c
4·5m
Ice plant
Key
a Marina centre building - cafe, dive centre, sailing and rowing centre, chandlery and showers
b Sea Life Centre
c Commercial berths - FVs & ferries
d Marina office

FENIT HARBOUR MARINA

Fenit Harbour & Marina
Fenit, Tralee, Co. Kerry, Republic of Ireland
Tel: +353 66 7136231 Fax: +353 66 7136473
Email: fenit.harbour@kerrycoco.ie www.kerrycoco.ie

VHF	Ch M
ACCESS	H24

Fenit Harbour Marina is tucked away in Tralee Bay, not far south of the Shannon Estuary. Besides offering a superb cruising ground, being within a day's sail of Dingle and Kilrush, the marina also provides a convenient base from which to visit inland attractions such as the picturesque tourist towns of Tralee and Killarney. This 120-berth marina accommodates boats up to 15m LOA and benefits from deep water at all states of the tide.

The small village of Fenit incorporates a grocery shop as well a several pubs and restaurants, while among the local activities are horse riding, swimming from one of the nearby sandy beaches and golfing.

FACILITIES AT A GLANCE
P
WC

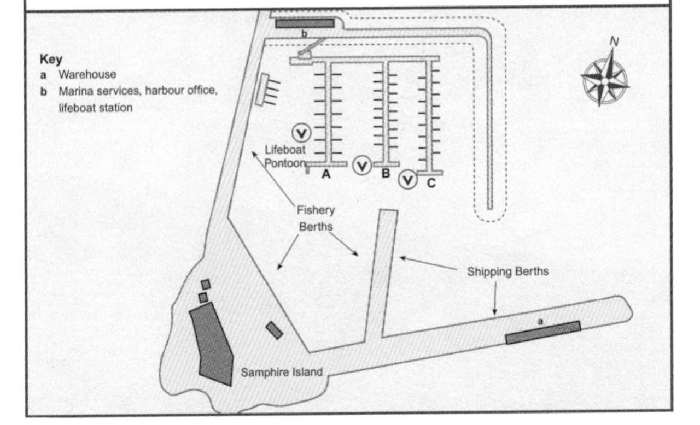
Key
a Warehouse
b Marina services, harbour office, lifeboat station
b
N
Lifeboat Pontoon
A
B
C
Fishery Berths
Shipping Berths
a
Samphire Island

KILRUSH MARINA

Kilrush Marina Ltd
Kilrush, Co. Clare, Ireland
Tel: +353 65 9052072 Mobile: +353 86 2313870
Email: info@kilrushmarina.ie

VHF	Ch 80
ACCESS	H24

Kilrush Marina and boatyard is well placed for exploring the unspoilt west coast of Ireland, including Galway Bay, Dingle, W Cork and Kerry. It also provides a gateway to over 150 miles of cruising on Lough Derg, the R Shannon and the Irish canal system. Accessed via lock gates, the marina lies at one end of the main street in Kilrush, the marina centre provides all the facilities for the visiting sailor. Kilrush is a vibrant market town with a long maritime history. A 15-minute ferry ride from the marina takes you to Scattery Is, once a 6th century monastic settlement but now only inhabited by wildlife. The Shannon Estuary is reputed for being the country's first marine Special Area of Conservation (SAC) and is home to Ireland's only known resident group of bottlenose dolphins.

FACILITIES AT A GLANCE
@
P
WC

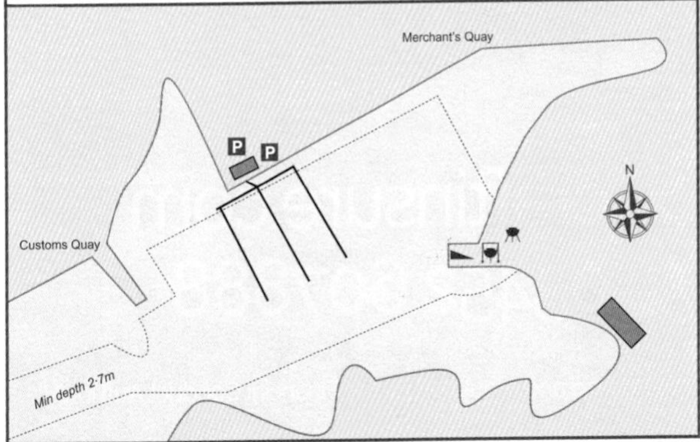
Merchant's Quay
P
P
N
Customs Quay
Min depth 2·7m

NORTH IRELAND – Liscannor Bay, clockwise to Lambay Island

Key to Marina Plans symbols

- Bottled gas
- Chandler
- Disabled facilities
- Electrical supply
- Electrical repairs
- Engine repairs
- First Aid
- Fresh Water
- Fuel - Diesel
- Fuel - Petrol
- Hardstanding/boatyard
- Internet Café
- Laundry facilities
- Lift-out facilities
- Parking
- Pub/Restaurant
- Pump out
- Rigging service
- Sail repairs
- Shipwright
- Shop/Supermarket
- Showers
- Slipway
- Toilets
- Telephone
- Trolleys
- Visitors berths
- Wi-Fi

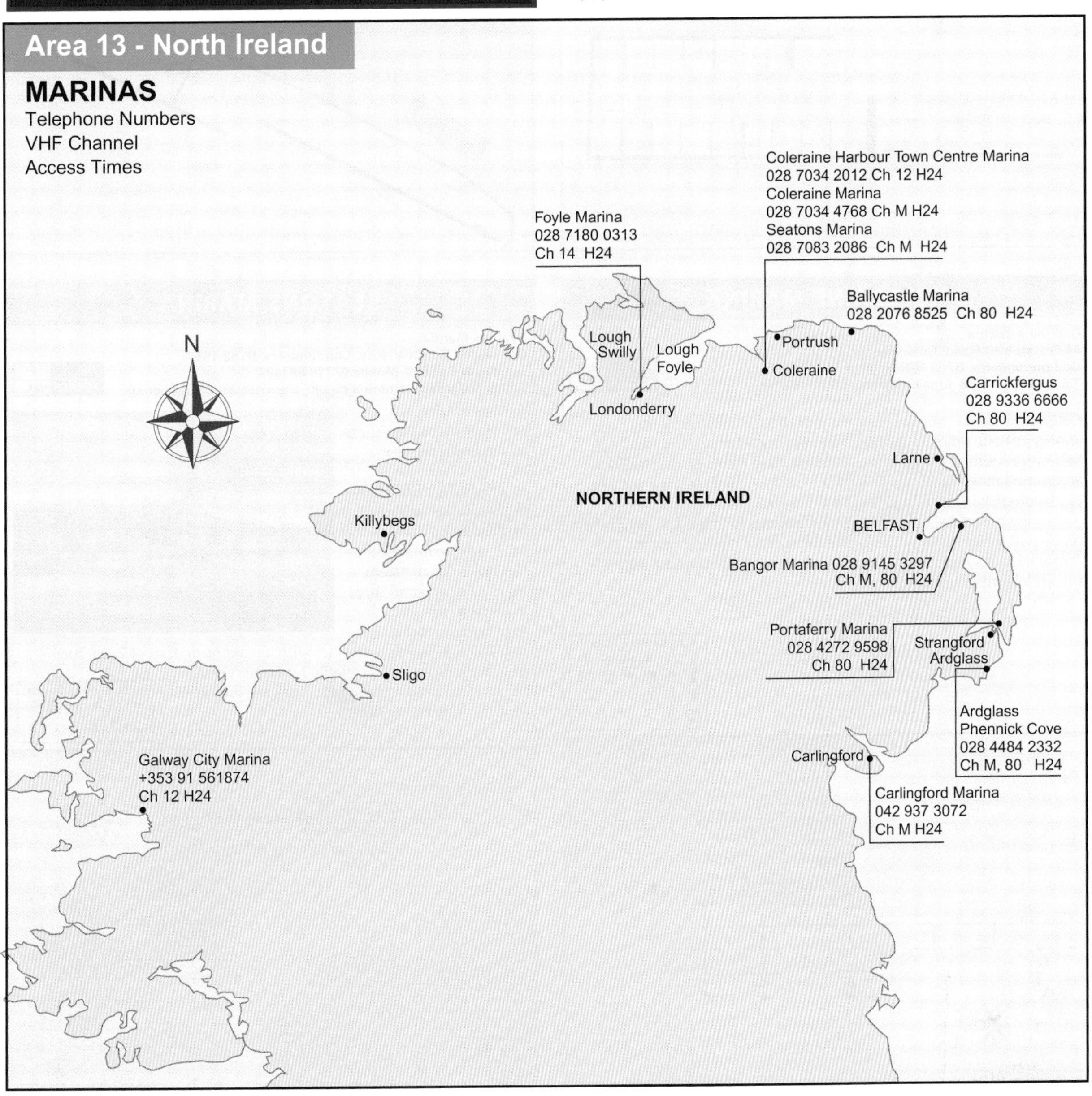
Area 13 - North Ireland
MARINAS
Telephone Numbers
VHF Channel
Access Times
Coleraine Harbour Town Centre Marina
028 7034 2012 Ch 12 H24
Coleraine Marina
028 7034 4768 Ch M H24
Seatons Marina
028 7083 2086 Ch M H24
Foyle Marina
028 7180 0313
Ch 14 H24
Ballycastle Marina
028 2076 8525 Ch 80 H24
N
Lough Swilly
Lough Foyle
Portrush
Coleraine
Londonderry
Carrickfergus
028 9336 6666
Ch 80 H24
Larne
NORTHERN IRELAND
Killybegs
BELFAST
Bangor Marina 028 9145 3297
Ch M, 80 H24
Portaferry Marina
028 4272 9598
Ch 80 H24
Strangford
Ardglass
Sligo
Ardglass
Phennick Cove
028 4484 2332
Ch M, 80 H24
Carlingford
Galway City Marina
+353 91 561874
Ch 12 H24
Carlingford Marina
042 937 3072
Ch M H24

GALWAY CITY MARINA

Galway City Marina
Galway Harbour Co, Harbour Office, Galway, Ireland
Tel: +353 91 561874 Fax: +353 91 563738
Email: info@theportofgalway.com

VHF	Ch 12
ACCESS	HW-2 to HW

The Galway harbour Company operates a small marina in the confines of Galway Harbour with an additional 60m of pontoon-walkway. Freshwater and electrical power is available at the pontoons. Power cars can be purchased from the harbour office during the day. A number of visitors pontoons are available for hire during the summer and for winter layup. Sailors intending to call to Galway Harbour should first make contact with the Harbour office to determine if a berth is available – advisable as demand is high in this quiet and beautiful part of Ireland.

FACILITIES AT A GLANCE

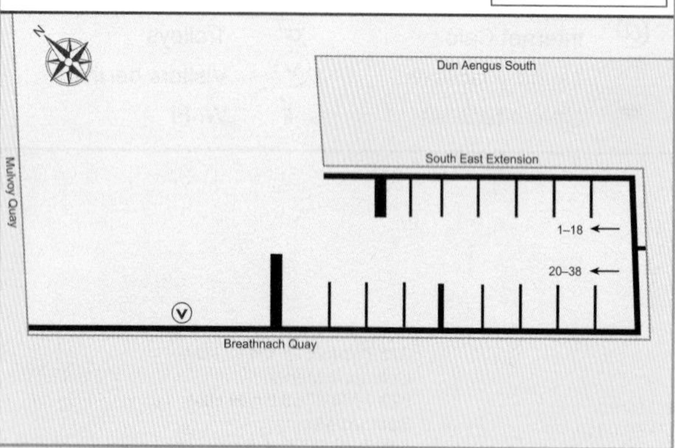
Dun Aengus South
South East Extension
Mulvoy Quay
1–18
20–38
Breathnach Quay

FOYLE MARINA

Foyle Marina
Londonderry Port, Lisahally, L'Derry, BT47 6FL
Tel: 02871 860555 Fax: 02871 861656
www.londonderryport.com/leisure
Email: info@londonderryport.com

VHF	Ch 14
ACCESS	H24

Foyle Marina lies in the heart of the city, 17M from the mouth of Lough Foyle, is accessible at any state of the tide and is sheltered from all directions of wind. Approach is via well-marked navigation channel with a maintained depth of 8m.

The marina has recently undergone extensive improvements with over 680m of secure pontoon mooring now available. Foyle Marina now offers full facilities to visiting vessels. Toilets and showers on site, water and electricity at each berth. Vessels up to 130m LOA can be accommodated. Craft can berth either side of the pontoons in depths of 5–7m at LW.

The pontoons are within easy walking distance of the city centre where you will find restaurants, bars, cinemas, shopping and a host of tourist attractions.

FACILITIES AT A GLANCE

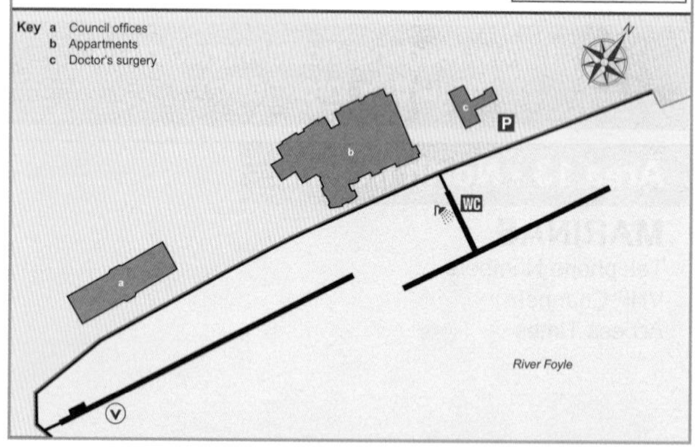
Key a Council offices
b Appartments
c Doctor's surgery
River Foyle

COLERAINE MARINA

Coleraine Marina
64 Portstewart Road, Coleraine,
Co Londonderry, BT52 1RR
Tel: 028 7034 4768 Email: rickiemac@talktalk.net

VHF	Ch M
ACCESS	H24

Coleraine Marina complex enjoys a superb location in sheltered waters just one mile north of the town of Coleraine and four and a half miles south of the River Bann Estuary and the open sea. Besides accommodating vessels up to 18m LOA, this modern marina with 78 berths offers hard standing, fuel and shower facilities.

Among one of the oldest known settlements in Ireland, Coleraine is renowned for its linen, whiskey and salmon. Its thriving commercial centre includes numerous shops, a four-screen cinema and a state-of-the-art leisure complex.

FACILITIES AT A GLANCE

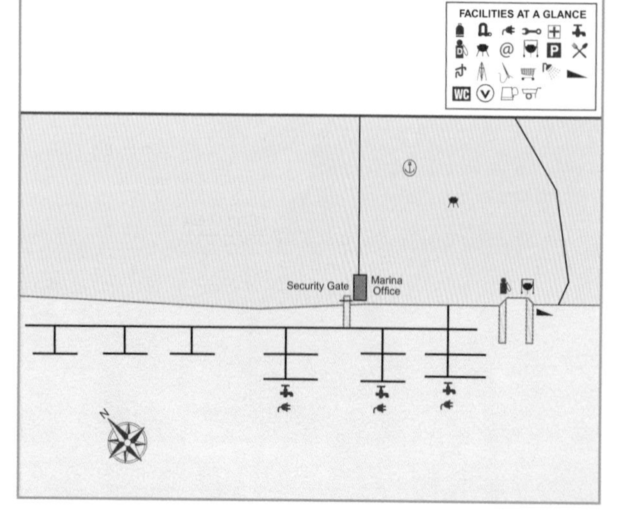
Security Gate
Marina Office

SEATONS MARINA

Seatons Marina
Drumslade Rd, Coleraine, Londonderry, BT52 1SE
Tel: 028 7083 2086 Mobile 07718 883099
Email: jill@seatonsmarina.co.uk www.seatonsmarina.co.uk

VHF	
ACCESS	H24

Seatons Marina is a privately owned business on the north coast of Ireland, which was established by Eric Seaton in 1962. It lies on the east bank of the River Bann, approximately two miles downstream from Coleraine and three miles from the sea. Long term pontoon berths are available for yachts up to 11.5 with a maximum draft of 2.4m; fore and aft moorings are available for larger vessels. Lift out and mast stepping facilities are provided by a 12 tonne trailer hoist.

FACILITIES AT A GLANCE

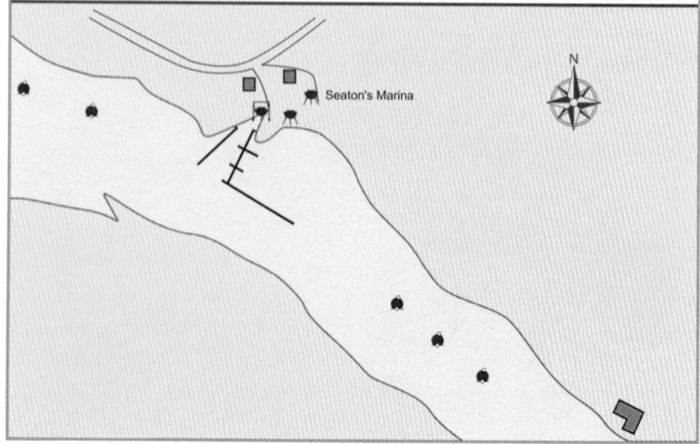
Seaton's Marina
N

COLERAINE HARBOUR MARINA

Coleraine Harbour Town Centre Marina
Coleraine Harbour Office, 4 Riversdale Road, Coleraine, BT52 1XA
Tel: 028 7034 2012 Mobile: 07742 242788
Email: info@coleraineharbour.com
www.coleraineharbour.com

VHF	Ch 12
ACCESS	H24

The Marina lies upstream about five miles from the sea. Ideally situated in a sheltered location in the centre of the town, just a few minutes stroll from a selection of shops, cafes, restaurants and bars. It is the ideal location for a short or long stay.

In addition to the pontoon berths there is a 40 tonne Roodberg slipway launch/ recovery trailer. Hard standing and covered storage are available.

FACILITIES AT A GLANCE

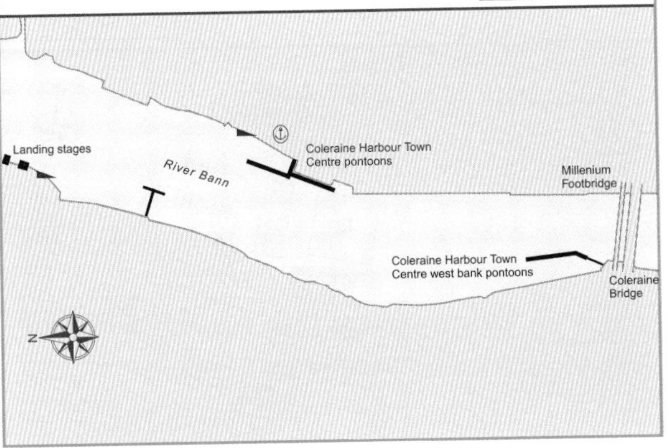
Landing stages
River Bann
Coleraine Harbour Town Centre pontoons
Millenium Footbridge
Coleraine Harbour Town Centre west bank pontoons
Coleraine Bridge
N

BALLYCASTLE MARINA

Ballycastle Marina
Bayview Road, Ballycastle, Northern Ireland
Tel: 028 2076 8525/07803 505084 Fax: 028 2076 6215
Email: info@moyle-council.org

VHF	Ch 80
ACCESS	H24

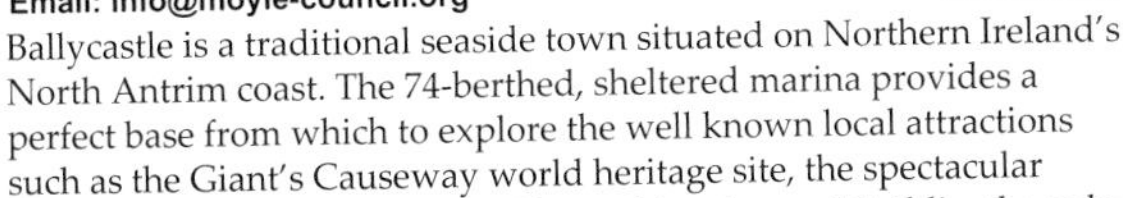

Ballycastle is a traditional seaside town situated on Northern Ireland's North Antrim coast. The 74-berthed, sheltered marina provides a perfect base from which to explore the well known local attractions such as the Giant's Causeway world heritage site, the spectacular Nine Glens of Antrim, and Rathlin, the only inhabited island in Northern Ireland. The most northern coastal marina in Ireland, Ballycastle is accessible at all states of the tide, although yachts are required to contact the marina on VHF Ch 80 before entering the harbour. Along the seafront are a selection of restaurants, bars and shops, while the town centre is only about a five minute walk away.

FACILITIES AT A GLANCE

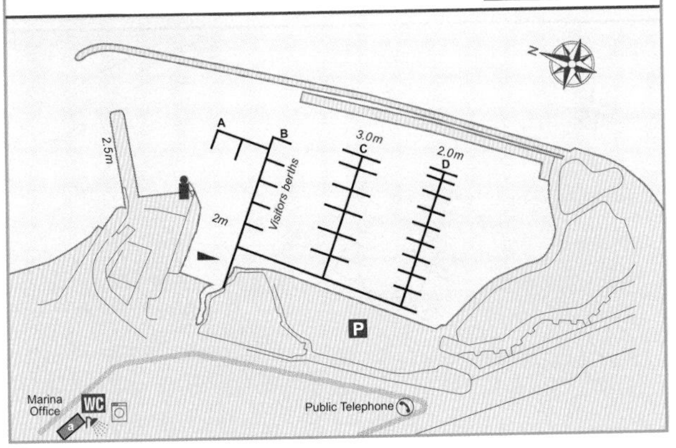
N
2.5m
A
B
3.0m
C
2.0m
D
2m
Visitors berths
P
Marina Office
WC
Public Telephone

CARRICKFERGUS MARINA

Carrickferus Marina
3 Quayside, Carrickfergus, Co. Antrim, BT38 8BJ
Tel: 028 9336 6666 Fax: 028 9335 0505
Email: marina.reception@midandeastantrim.gov.uk
www.midandeastantrim.gov.uk

VHF	Ch 80
ACCESS	H24

Carrickfergus Marina is situated on the north shore of Belfast Lough overlooked by the town's medieval 12th century Norman Castle. Known as the gateway to the Causeway Coastal Route, Carrickfergus is also well connected to main arterial routes headed to Belfast (7 miles) and further afield.

Our 300-berth, fully serviced marina has earned the prestigious 5 Gold Anchor Award and European Blue Flag status. No detail is overlooked in our exclusive berth holder facilities. From immaculately presented showers and personal laundry service, together with bespoke marine services, these are just some of the facilities available during a stay.

FACILITIES AT A GLANCE

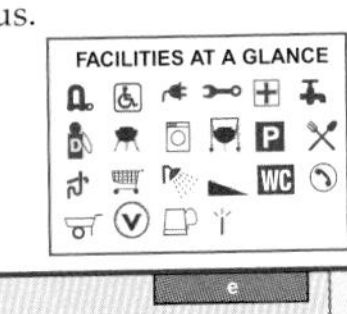

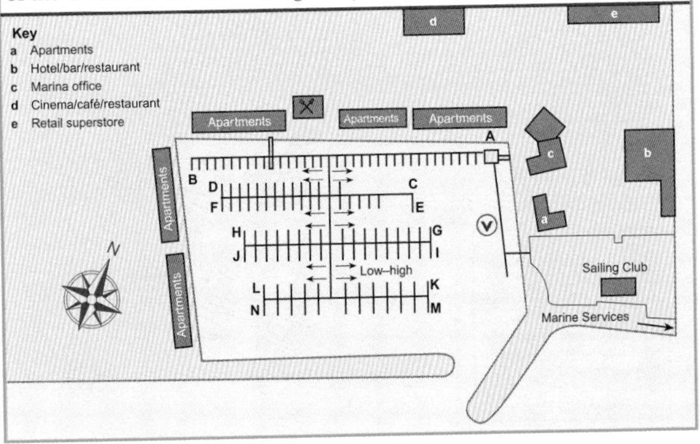
Key
a Apartments
b Hotel/bar/restaurant
c Marina office
d Cinema/café/restaurant
e Retail superstore
Apartments
Apartments
Apartments
A
B
C
D
E
F
G
H
I
J
K
L
M
N
Low–high
Sailing Club
Marine Services
N

BANGOR MARINA

Bangor Marina
Bangor, Co. Down, BT20 5ED
Tel: 028 9145 3297 Fax: 028 9145 3450
Email: bangor@boatfolk.co.uk
www.boatfolk.co.uk/bangormarina

VHF	Ch 11, 80
ACCESS	H24

Situated on the south shore of Belfast Lough, Bangor is located close to the Irish Sea cruising routes. The marina is right at the town's centre, within walking distance of shops, restaurants, hotels and bars. The Tourist information centre is across the road from marina reception and there are numerous visitors' attractions in the Borough. The Royal Ulster Yacht Club and the Ballyholme Yacht Club are both nearby and welcome visitors.

FACILITIES AT A GLANCE

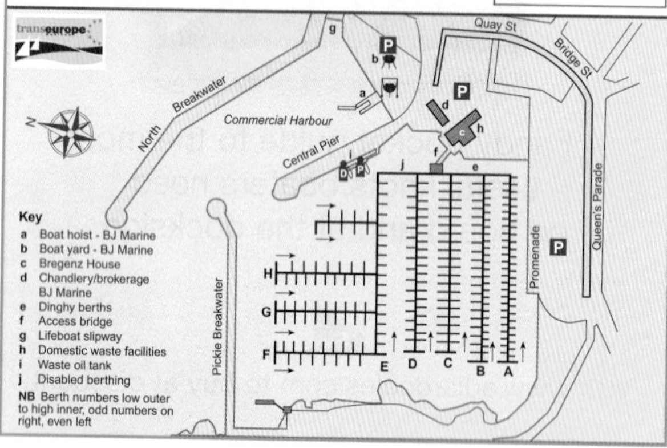
transeurope
Quay St
Bridge St
North Breakwater
Commercial Harbour
Central Pier
Queen's Parade
Promenade
Pickie Breakwater
H
G
F
E
D
C
B
A
Key
a Boat hoist - BJ Marine
b Boat yard - BJ Marine
c Bregenz House
d Chandlery/brokerage BJ Marine
e Dinghy berths
f Access bridge
g Lifeboat slipway
h Domestic waste facilities
i Waste oil tank
j Disabled berthing
NB Berth numbers low outer to high inner, odd numbers on right, even left

CARLINGFORD MARINA

Carlingford Marina
Co. Louth, Ireland
Tel: +353 (0)42 937 3072 Fax: +353 (0)42 937 3075
Email: info@carlingfordmarina.ie
www.carlingfordmarina.ie

VHF	Ch M
ACCESS	H24

Carlingford Lough is an eight-mile sheltered haven between the Cooley Mountains to the south and the Mourne Mountains to the north. The marina is situated on the southern shore, about four miles from Haulbowline Lighthouse, and can be easily reached via a deep water shipping channel. Among the most attractive destinations in the Irish Sea, Carlingford is only 60 miles from the Isle of Man and within a day's sail from Strangford Lough and Ardglass. Full facilities in the marina include a first class bar and restaurant offering superb views across the water.

FACILITIES AT A GLANCE

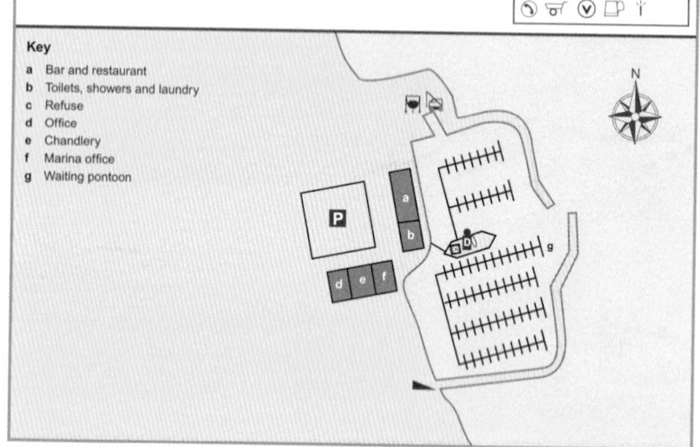
Key
a Bar and restaurant
b Toilets, showers and laundry
c Refuse
d Office
e Chandlery
f Marina office
g Waiting pontoon
N

ARDGLASS MARINA

Ardglass Marina
19 Quay Street, Ardglass, BT30 7SA
Tel: 028 4484 2332
Email: infoardglassmarina@gmail.com
www.ardglassmarina.co.uk

VHF	Ch M, 80
ACCESS	H24

Situated just south of Strangford, Ardglass has the capacity to accommodate up to 22 yachts as well as space for small craft. Despite being relatively small in size, the marina boasts an extensive array of facilities, either on site or close at hand. Recent access improvements have been made for wheelchair users in the shower/WC and reception. Grocery stores, a post office, chemist and off-licence, are all within a five-minute walk from the marina. Among the local onshore activities are golf, mountain climbing in Newcastle, which is 18 miles south, as well as scenic walks at Ardglass and Delamont Park.

FACILITIES AT A GLANCE

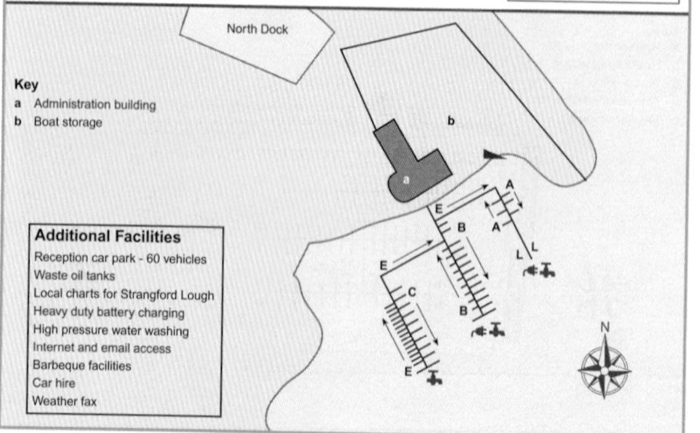
North Dock
Key
a Administration building
b Boat storage
Additional Facilities
Reception car park - 60 vehicles
Waste oil tanks
Local charts for Strangford Lough
Heavy duty battery charging
High pressure water washing
Internet and email access
Barbeque facilities
Car hire
Weather fax
N

PORTAFERRY MARINA

Portaferry Marina
1 Mill View, Portaferry, BT22 1LQ
Mobile: 07703 209780 Fax: 028 4272 9784
Email: info@portaferrymarina.co.uk

VHF	Ch 80
ACCESS	H24

Portaferry Marina lies on the east shore of the Narrows, the gateway to Strangford Lough on the north east coast of Ireland. A marine nature reserve of outstanding natural beauty, the Lough offers plenty of recreational activities.

The marina, which caters for draughts of up to 2.5m, is fairly small, accommodating around 30 yachts. The office is situated about 200m from the marina itself, where you will find ablution facilities along with a launderette.

Portaferry incorporates several pubs and restaurants as well as a few convenience stores, while one of its prime attractions is the Exploris Aquarium. Places of historic interest in the vicinity include Castleward, an 18th century mansion in Strangford, and Mount Stewart House & Garden in Newtownards.

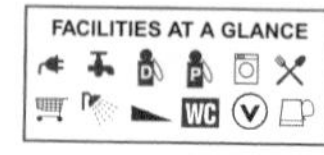
FACILITIES AT A GLANCE

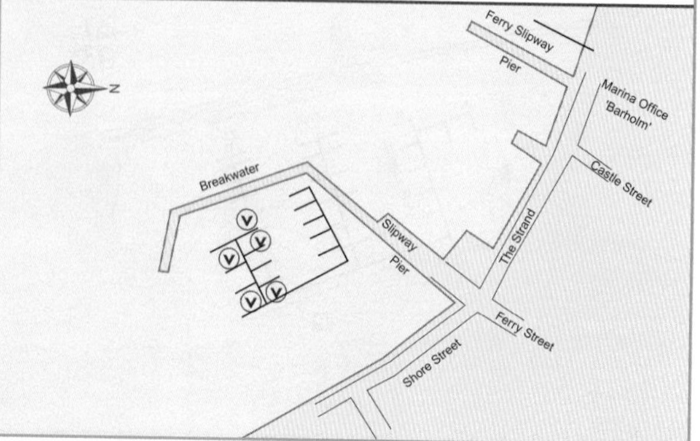
Ferry Slipway
Pier
Marina Office 'Barholm'
Castle Street
The Strand
Breakwater
Slipway
Pier
Ferry Street
Shore Street
N

CHANNEL ISLANDS – Guernsey & Jersey

Key to Marina Plans symbols

- Bottled gas
- Chandler
- Disabled facilities
- Electrical supply
- Electrical repairs
- Engine repairs
- First Aid
- Fresh Water
- Fuel - Diesel
- Fuel - Petrol
- Hardstanding/boatyard
- Internet Café
- Laundry facilities
- Lift-out facilities
- Parking
- Pub/Restaurant
- Pump out
- Rigging service
- Sail repairs
- Shipwright
- Shop/Supermarket
- Showers
- Slipway
- Toilets
- Telephone
- Trolleys
- Visitors berths
- Wi-Fi

Area 14 - Channel Islands

MARINAS
Telephone Numbers
VHF Channel
Access Times

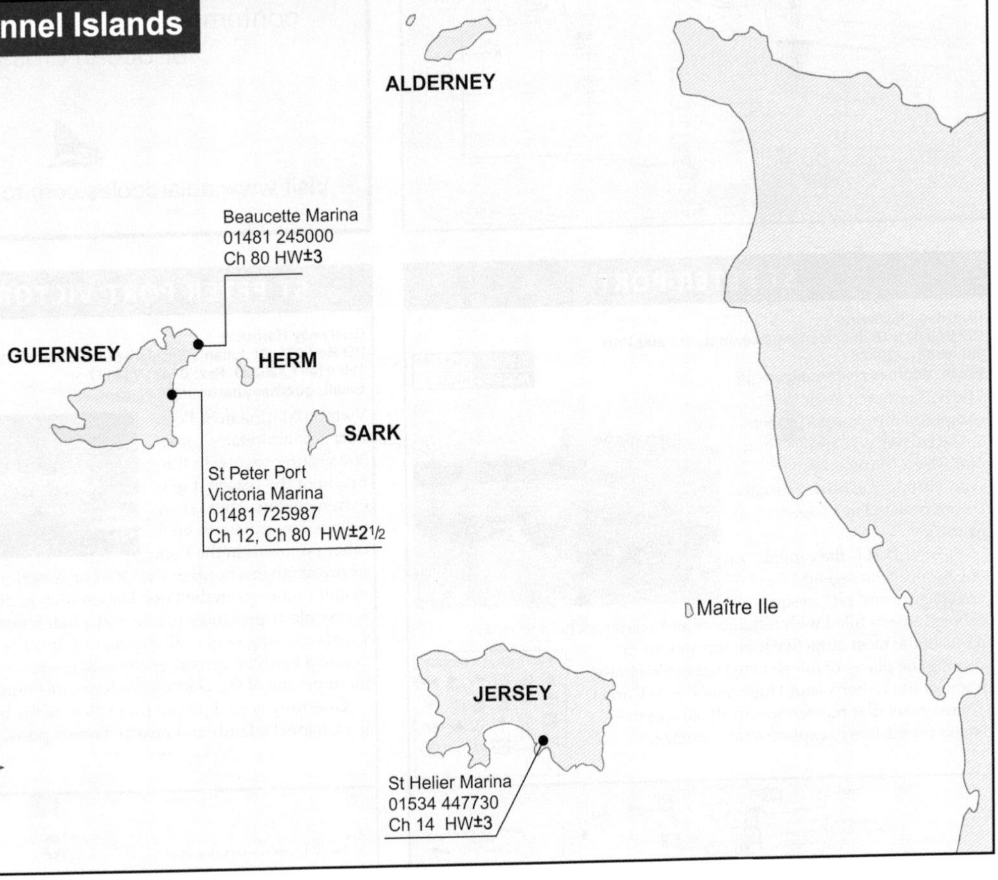

ALDERNEY

Beaucette Marina
01481 245000
Ch 80 HW±3

GUERNSEY

HERM

SARK

St Peter Port
Victoria Marina
01481 725987
Ch 12, Ch 80 HW±2½

Maître Ile

N

JERSEY

St Helier Marina
01534 447730
Ch 14 HW±3

BEAUCETTE MARINA

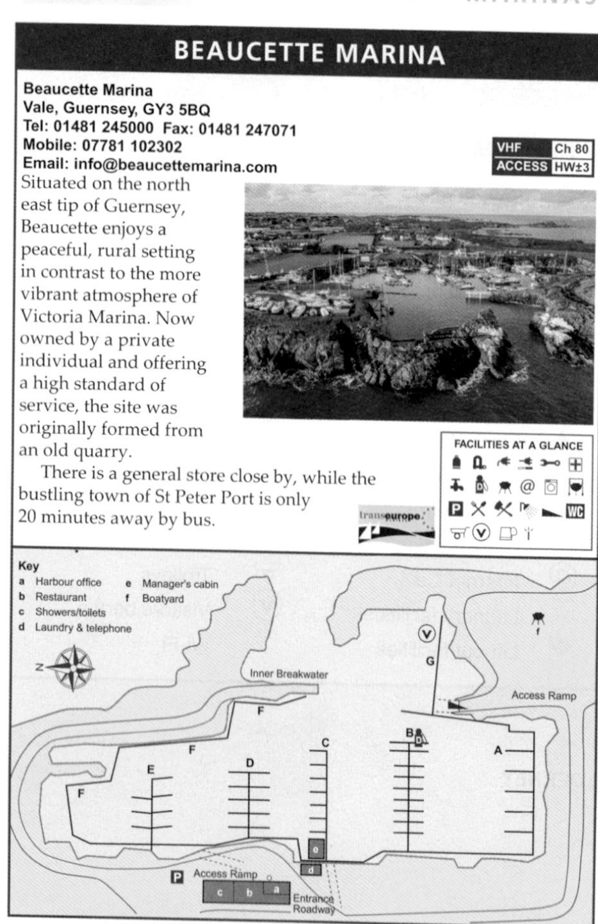

Beaucette Marina
Vale, Guernsey, GY3 5BQ
Tel: 01481 245000 Fax: 01481 247071
Mobile: 07781 102302
Email: info@beaucettemarina.com

VHF	Ch 80
ACCESS	HW±3

Situated on the north east tip of Guernsey, Beaucette enjoys a peaceful, rural setting in contrast to the more vibrant atmosphere of Victoria Marina. Now owned by a private individual and offering a high standard of service, the site was originally formed from an old quarry.

There is a general store close by, while the bustling town of St Peter Port is only 20 minutes away by bus.

FACILITIES AT A GLANCE
transeurope
Key
a Harbour office
b Restaurant
c Showers/toilets
d Laundry & telephone
e Manager's cabin
f Boatyard
Inner Breakwater
Access Ramp
Access Ramp
Entrance Roadway

REEDS OCEAN HANDBOOK
THE COMPREHENSIVE POCKET GUIDE
BILL JOHNSON
The ideal primer for anyone contemplating an offshore passage or ocean crossing
Visit www.adlardcoles.com to buy at discount

ST PETER PORT

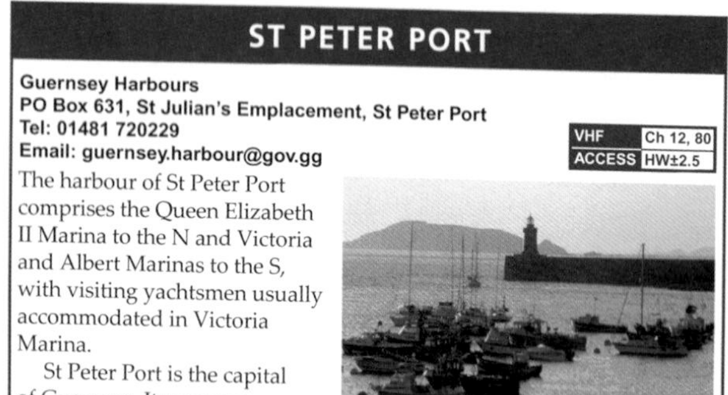

Guernsey Harbours
PO Box 631, St Julian's Emplacement, St Peter Port
Tel: 01481 720229
Email: guernsey.harbour@gov.gg

VHF	Ch 12, 80
ACCESS	HW±2.5

The harbour of St Peter Port comprises the Queen Elizabeth II Marina to the N and Victoria and Albert Marinas to the S, with visiting yachtsmen usually accommodated in Victoria Marina.

St Peter Port is the capital of Guernsey. Its regency architecture and picturesque cobbled streets filled with restaurants and boutiques help to make it one of the most attractive harbours in Europe. Among the places of interest are Hauteville House, home of the writer Victor Hugo, and Castle Cornet. There are regular bus services to all parts of the island for visitors to explore a rich heritage.

FACILITIES AT A GLANCE

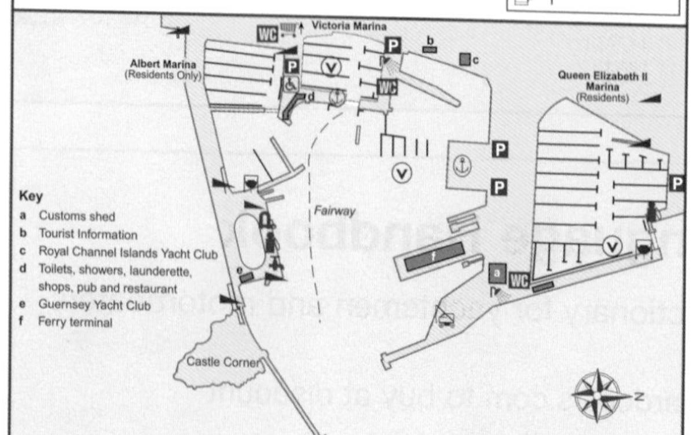

Victoria Marina
Albert Marina (Residents Only)
Queen Elizabeth II Marina (Residents)
Fairway
Castle Cornet
Key
a Customs shed
b Tourist Information
c Royal Channel Islands Yacht Club
d Toilets, showers, launderette, shops, pub and restaurant
e Guernsey Yacht Club
f Ferry terminal

ST PETER PORT VICTORIA MARINA

Guernsey Harbours
PO Box 631, St Julian's Emplacement, St Peter Port
Tel: 01481 720229 Fax: 01481 714177
Email: guernsey.harbour@gov.gg

VHF	Ch 80
ACCESS	HW±2.5

Victoria Marina in St Peter Port accommodates some 300 visiting yachts. In the height of the season it gets extremely busy, but when full visitors can berth on 5 other pontoons in the Pool or pre-arrange a berth in the QE II or Albert marinas. There are no visitor moorings in the Pool. Depending on draught, the marina is accessible approximately two and a half hours either side of HW, with yachts crossing over a sill drying to 4.2m. The marina dory will direct you to a berth on arrival or else will instruct you to moor on one of the waiting pontoons just outside.

Guernsey is well placed for exploring the rest of the Channel Islands and nearby French ports.

FACILITIES AT A GLANCE

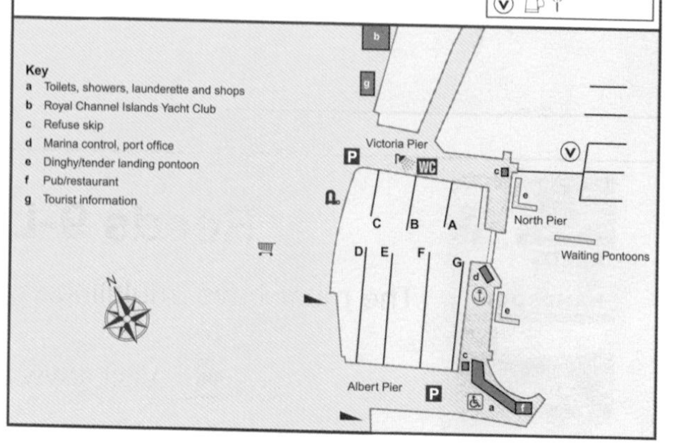

Key
a Toilets, showers, launderette and shops
b Royal Channel Islands Yacht Club
c Refuse skip
d Marina control, port office
e Dinghy/tender landing pontoon
f Pub/restaurant
g Tourist information
Victoria Pier
North Pier
Waiting Pontoons
Albert Pier

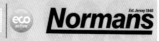

ST HELIER HARBOUR

St Helier Harbour
Maritime House, La Route du Port Elizabeth
St Helier, Jersey, JE1 1HB
Tel: 01534 447708
www.portofjersey.je Email: marinas@ports.je

VHF	Ch 14
ACCESS	HW±3

Set in the Norman Breton Gulf, Jersey is the most southerly of the Channel Islands, offering over 200 visitor berths on a flexible daily, weekly or monthly basis. With its close proximity to the adjacent French coast and with sheltered bays and anchorages there are plenty of opportunities to explore new cruising areas, making Jersey an ideal base. Jersey airport is only 15 minutes away from St Helier Marina by bus or car.

Visiting craft are directed into St Helier Marina which is located in the town with access is HW±3. Alternatively, there is a holding pontoon just outside of the marina entrance providing full services and walk ashore access. Visiting vessels of up to 20m may be directed into Elizabeth Marina by prior arrangement.

FACILITIES AT A GLANCE

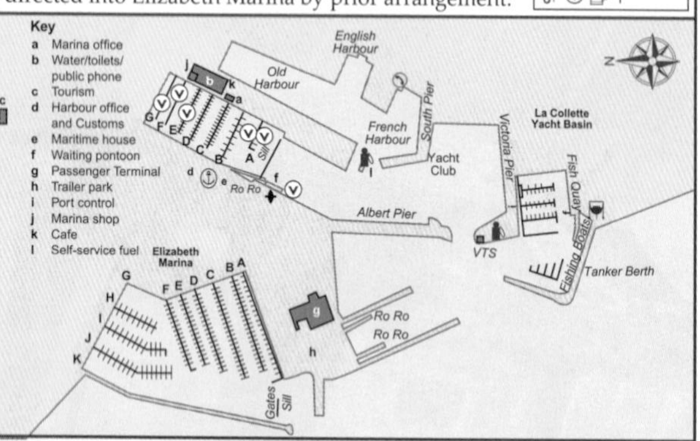
Key
a Marina office
b Water/toilets/public phone
c Tourism
d Harbour office and Customs
e Maritime house
f Waiting pontoon
g Passenger Terminal
h Trailer park
i Port control
j Marina shop
k Cafe
l Self-service fuel
English Harbour
Old Harbour
French Harbour
South Pier
Yacht Club
Victoria Pier
La Collette Yacht Basin
Fish Quay
Fishing Boats
Tanker Berth
Albert Pier
VTS
Ro Ro
Elizabeth Marina
Gates
Sill

SECTION 2
MARINE SUPPLIES AND SERVICES GUIDE

ADHESIVES

Casco Adhesives
Darwen 07710 546899

CC Marine Services Ltd
West Mersea 07751 734510

Industrial Self Adhesives Ltd
Nottingham 0115 9681895

Sika Ltd Garden City 01707 394444

Technix Rubber & Plastics Ltd
Southampton 01489 789944

Tiflex Liskeard 01579 320808

Trade Grade Products Ltd
Poole 01202 820177

UK Epoxy Resins
Burscough 01704 892364

Wessex Resins & Adhesives Ltd
Romsey 01794 521111

3M United Kingdom plc
Bracknell 01344 858315

ASSOCIATIONS/ AGENCIES

Cruising Association
London 020 7537 2828

Fishermans Mutual Association (Eyemouth) Ltd
Eyemouth 01890 750373

Maritime and Coastguard Agency
Southampton 0870 6006505

Royal Institute of Navigation
London 020 7591 3130

Royal National Lifeboat Institution
Poole 01202 663000

Royal Yachting Association (RYA) Southampton 0845 345 0400

BERTHS & MOORINGS

ABC Powermarine
Beaumaris 01248 811413

Aqua Bell Ltd Norwich 01603 713013

Ardfern Yacht Centre
Lochgilphead 01852 500247/500636

Ardmair Boat Centre
Ullapool 01854 612054

Arisaig Marine Ltd
Inverness-shire 01687 450224

Bristol Boat Ltd Bristol 01225 872032

British Waterways
Argyll 01546 603210

Burgh Castle Marine
Norfolk 01493 780331

Cambrian Marine Services Ltd
Cardiff 029 2034 3459

Chelsea Harbour Ltd
London 020 7225 9108

Clapson & Son (Shipbuilders) Ltd
Barton-on-Humber 01652 635620

Crinan Boatyard, Crinan 01546 830232

Dartside Quay Brixham 01803 845445

Douglas Marine
Preston 01772 812462

Dublin City Moorings
Dublin +353 1 8183300

Emsworth Yacht Harbour
Emsworth 01243 377727

Exeter Ship Canal 01392 274306

HAFAN PWLLHELI
Glan Don, Pwllheli, Gwynedd LL53 5YT
Tel: (01758) 701219
Fax: (01758) 701443 VHF Ch80
Hafan Pwllheli has over 400 pontoon berths and offers access at virtually all states of the tide. Ashore, its modern purpose-built facilities include luxury toilets, showers, launderette, a secure boat park for winter storage, 40-ton travel hoist, mobile crane and plenty of space for car parking. Open 24-hours a day, 7 days a week.

Highway Marine
Sandwich 01304 613925

Iron Wharf Boatyard
Faversham 01795 536296

Jalsea Marine Services Ltd
Northwich 01606 77870

Jersey Harbours
St Helier 01534 447788

Jones (Boatbuilders), David
Chester 01244 390363

Lawrenny Yacht Station
Kilgetty 01646 651212

MacFarlane & Son
Glasgow 01360 870214

NEPTUNE MARINA LTD
Neptune Quay, Ipswich, Suffolk IP4 1AX
Tel: (01473) 215204
Fax: (01473) 215206
e-mail: enquiries@neptune-marina.com
www.neptune-marina.com
The quay to the heart of Ipswich! Call Ipswich lock gates on Channel 68 and Neptune Marina on Channel 80. Bring your crew to the wonderful Ipswich waterfront, with plenty of watering holes and town centre activities. You won't want to leave!

Orkney Marinas Ltd
Kirkwall 07810 465835

V Marine
Shoreham-by-Sea 01273 461491

Sutton Harbour Marina
Plymouth 01752 204186

WicorMarine Fareham 01329 237112

Winters Marine Ltd
Salcombe 01548 843580

Yarmouth Marine Service
Yarmouth 01983 760521

Youngboats
Faversham 01795 536176

BOAT BUILDERS & REPAIRS

ABC Hayling Island 023 9246 1968

ABC Powermarine
Beaumaris 01248 811413

Advance Yacht Systems
Southampton 023 8033 7722

Aqua-Star Ltd
St Sampsons 01481 244550

Ardoran Marine
Oban 01631 566123

Baumbach Bros Boatbuilders
Hayle 01736 753228

Beacon Boatyard
Rochester 01634 841320

Bedwell & Co
Walton on the Naze 01255 675873

Blackwell, Craig
Co Meath +353 87 677 9605

Boyd Boat Building
Falmouth 07885 436722

Boatcraft
Ardrossan 01294 603047

B+ St Peter Port 01481 726071

Brennan, John
Dun Laoghaire +353 1 280 5308

Burghead Boat Centre
Findhorn 01309 690099

Carrick Marine Projects
Co Antrim 02893 355884

Chapman & Hewitt Boatbuilders
Wadebridge 01208 813487

Chicks Marine Ltd
Guernsey 01481 723716

Clarence Boatyard
East Cowes 01983 294243

Cooks Maritime Craftsmen - Poliglow
Lymington 01590 675521

Creekside Boatyard (Old Mill Creek)
Dartmouth 01803 832649

CTC Marine & Leisure
Middlesbrough 01642 372600

Davies Marine Services
Ramsgate 01843 586172

Dickie International
Bangor 01248 363400

Dickie International
Pwllheli 01758 701828

East Llanion Marine Ltd
Pembroke Dock 01646 686866

Emblem Enterprises
East Cowes 01983 294243

Fairlie Quay
Fairlie 01475 568267

Fairweather Marine
Fareham 01329 283500

Farrow & Chambers Yacht Builders
Humberston
www.farrowandchambers.co.uk

Fast Tack Plymouth 01752 255171

Fergulsea Engineering
Ayr 01292 262978
Ferrypoint Boat Co
Youghal +353 24 94232
Floetree Ltd (Loch Lomond Marina)
Balloch 01389 752069
Freshwater Boatyard
Truro 01326 270443
Frogmore Boatyard
Kingsbridge 01548 531257
Furniss Boat Building
Falmouth 01326 311766
Gallichan Marine Ltd
Jersey 01534 746387
Garvel Clyde
Greenock 01475 725372
Goodchild Marine Services
Great Yarmouth 01493 782301
Gosport Boatyard
Gosport 023 9252 6534
Gweek Quay Boatyard
Helston 01326 221657
Halls
Walton on the Naze 01255 675596
Harris Pye Marine
Barry 01446 720066
Haven Boatyard
Lymington 01590 677073
Hayling Yacht Company
Hayling Island 023 9246 3592
Hoare Ltd, Bob
Poole 01202 736704
Holyhead Boatyard
Holyhead 01407 760111
Jackson Marine
Lowestoft 01502 539772
Jackson Yacht Services
Jersey 01534 743819
JEP Marine
Canterbury 01227 710102
JWS Marine Services
Southsea 023 9275 5155
Kimelford Yacht Haven
Oban 01852 200248
Kingfisher Marine
Weymouth 01305 766595
Kingfisher Ultraclean UK Ltd
Tarporley 0800 085 7039
King's Boatyard
Pin Mill 01473 780258
Kinsale Boatyard
Kinsale +353 21477 4774
Kippford Slipway Ltd
Dalbeattie 01556 620249
Lawrenny Yacht Station
Lawrenny 01646 651212
Lencraft Boats Ltd
Dungarvan +353 58 682220

Mackay Boatbuilders
Arbroath 01241 872879
Marine Blast
Holy Loch 01369 705394
Marine Services
Norwich 01692 582239
Mashford Brothers
Torpoint 01752 822232
Mayor & Co Ltd, J
Preston 01772 812250
Mears, HJ Axmouth 01297 23344
Mill, Dan, Galway +353 86 337 9304
Miller Marine
Tyne & Wear 01207 542149
Morrison, A
Killyleagh 028 44828215
Moss (Boatbuilders), David
Thornton-Cleveleys 01253 893830
Multi Marine Composites Ltd
Torpoint 01752 823513
Newing, Roy E
Canterbury 01227 860345
Noble and Sons, Alexander
Girvan 01465 712223
Northney Marine Services
Hayling Island 023 9246 9246
Northshore Sport & Leisure
King's Lynn 01485 210236
O'Sullivans Marine Ltd
Tralee +353 66 7124957
Pachol, Terry
Brighton 01273 682724
Partington Marine Ltd, William
Pwllheli 01758 612808
Pasco's Boatyard
Truro 01326 270269
Penrhos Marine
Aberdovey 01654 767478
Penzance Marine Services
Penzance 01736 361081
PJ Bespoke Boat Fitters Ltd
Crewe 01270 812244
Preston Marine Services Ltd
Preston 01772 733595
Red Bay Boats Ltd
Cushendall 028 2177 1331
Reliance Marine
Wirral 0151 625 5219
Reever, Adrian
Maldon 07548 744090
Retreat Boatyard Ltd
Exeter 01392 874720/875934
Richardson Boatbuilders, Ian
Stromness 01856 850321
Richardson Yacht Services Ltd
Newport 01983 821095

Roberts Marine Ltd, S
Liverpool 0151 707 8300
Rothman Pantall & Co
Fareham 01329 280221
Rustler Yachts
Falmouth 01326 310210
Salterns Boatyard
Poole 01202 707391
Sea & Shore Ship Chandler
Dundee 01382 450666
Seamark-Nunn & Co
Felixstowe 01394 275327
Seapower
Woolverstone 01473 780090
Slipway Cooperative Ltd
Bristol 0117 907 9938
Small, Donal
Galway +353 83 1831057
Smith, GB, & Sons
Rock 01208 862815
Spicer Boatbuilder, Nick
Weymouth Marina 01305 767118
Squadron Marine Ltd
Poole 01202 674531
Storrar Marine Store
Newcastle upon Tyne 0191 266 1037
TT Marine Ashwell 01462 742449
Waterfront Marine
Bangor 01248 352513
Way, A&R, Boat Building
Tarbert, Loch Fyne 01546 606657
WestCoast Marine
Troon 01292 318121
Western Marine
Dublin +353 1 280 0321
Wigmore Wright Marine Services
Penarth 029 2070 9983
Williams, Peter
Fowey 01726 870987
WQI Ltd
Bournemouth 01202 771292
Yarmouth Marine Service
01983 760521
Youngboats
Faversham 01795 536176

BOATYARD SERVICES & SUPPLIES

ABC Marine
Hayling Island 023 9246 1968
Abersoch Boatyard Services Ltd
Abersoch 01758 713900
Amble Boat Co Ltd
Amble 01665 710267
Amsbrisbeg Ltd
Port Bannatyne 01700 831215
Ardmair Boat Centre
Ullapool 01854 612054

Ardmaleish Boat Building Co Rothesay 01700 502007
www.ardoran.co.uk
W coast Scotland. All marine facilities.
Ardrishaig Boatyard
Lochgilphead 01546 603280
Arklow Slipway
Arklow +353 402 33233
Baltic Wharf Boatyard
Totnes 01803 867922
Baltimore Boatyard
Baltimore +353 28 20444
Bates, Declan
Kilmore Quay +353 87 252 9936
Bedwell and Co
Walton-on-the-Naze 01255 675873
Berthon Boat Co
Lymington 01590 673312
Birdham Shipyard
Chichester 01243 512310
BJ Marine Ltd Bangor 028 91271434
Blagdon, A
Plymouth 01752 561830
Boatcraft Ardrossan 01294 603047
Boatworks + Ltd
St Peter Port 01481 726071
Brennan, John
Dun Laoghaire +353 1 280 5308
Brighton Marina Boatyard
Brighton 01273 819919
Bristol Marina (Yard)
Bristol 0117 921 3198
Buckie Shipyard Ltd
Buckie 01542 831245
Bucklers Hard Boat Builders Ltd
Brockenhurst 01590 616214
C & J Marine Services
Newcastle Upon Tyne 0191 295 0072
Caley Marina Inverness 01463 236539
Cambrian Boat Centre
Swansea 01792 655925
Cambrian Marine Services Ltd
Cardiff 029 2034 3459
Cantell and Son Ltd
Newhaven 01273 514118
Canvey Yacht Builders Ltd
Canvey Island 01268 696094
Carroll's Ballyhack Boatyard
New Ross +353 51 389164
Castlepoint Boatyard
Crosshaven +353 21 4832154
Chabot, Gary
Newhaven 07702 006767
Chapman & Hewitt Boatbuilders
Wadebridge 01208 813487
Chippendale Craft Rye 01797 227707
Clapson & Son (Shipbuilders) Ltd
Barton on Humber 01652 635620
Clarence Boatyard
East Cowes 01983 294243

Coastal Marine Boatbuilders
Eyemouth 01890 750328
Coastcraft Ltd
Cockenzie 01875 812150
Coates Marine Ltd
Whitby 01947 604486
Connor, Richard
Coleraine 07712 115751
Coombes, AA
Bembridge 01983 872296
Corpach Boatbuilding Company
Fort William 01397 772861
Craobh Marina
By Lochgilphead 01852 500222
Creekside Boatyard (Old Mill Creek)
Dartmouth 01803 832649
Crinan Boatyard
By Lochgilphead 01546 830232
Crosshaven Boatyard Co Ltd
Crosshaven +353 21 831161
Dale Sailing Co Ltd
Neyland 01646 603110
Darthaven Marina
Kingswear 01803 752242
Dartside Quay
Brixham 01803 845445
Dauntless Boatyard Ltd
Canvey Island 01268 793782
Davis's Boatyard Poole 01202 674349
Dinas Boat Yard Ltd
Y Felinheli 01248 671642
Dorset Lake Shipyard Ltd
Poole 01202 674531
Dorset Yachts Co Ltd
Poole 01202 674531
Douglas Boatyard
Preston 01772 812462
Dover Yacht Co Dover 01304 201073
Dun Laoghaire Marina
Dun Laoghaire +353 1 2020040
Elephant Boatyard
Southampton 023 8040 3268
Elton Boatbuilding Ltd
Kirkcudbright 01557 330177
Felixstowe Ferry Boatyard
Felixstowe 01394 282173
Ferguson Engineering
Wexford +353 6568 66822133
Ferry Marine South
Queensferry 0131 331 1233
Findhorn Boatyard
Findhorn 01309 690099
Firmhelm Ltd Pwllheli 01758 612251
Fowey Boatyard
Fowey 01726 832194
Fox's Marina Ipswich 01473 689111
Frank Halls & Son
Walton on the Naze 01255 675596

Freeport Marine
Jersey 01534 888100
Furniss Boat Building
Falmouth 01326 311766
Garval Clyde
Greenock 01475 725372
Goodchild Marine Services
Great Yarmouth 01493 782301
Gosport Boatyard
Gosport 023 9252 6534
Gweek Quay Boatyard
Helston 01326 221657
Haines Boatyard
Chichester 01243 512228
Harbour Marine
Plymouth 01752 204691
Harbour Marine Services Ltd
Southwold 01502 724721
Harris Pye Marine Barry 01446 720066
Hartlepool Marine Engineering
Hartlepool 01429 867883
Hayles, Harold
Yarmouth, IoW 01983 760373
Henderson, J Shiskine 01770 860259
Heron Marine
Whitstable 01227 361255
Hewitt, George
Binham 01328 830078
Holyhead Marina & Trinity Marine Ltd
Holyhead 01407 764242
Instow Marine Services
Bideford 01271 861081
Ipswich Haven Marina
Ipswich 01473 236644
Iron Wharf Boatyard
Faversham 01795 536296
Island Boat Services
Port of St Mary 01624 832073
Isle of Skye Yachts
Ardvasar 01471 844216
Jalsea Marine Services Ltd Weaver Shipyard, Northwich 01606 77870
JBS Group
Peterhead 01779 475395
J B Timber Ltd
North Ferriby 01482 631765
Jersey Harbours Dept
St Helier 01534 885588
Kilnsale Boatyard
Kinsale +353 21 4774774
Kingfisher Ultraclean UK Ltd
Tarporley 01928 787878
Kilrush Marina Boatyard
Kilrush +35 87 7990091
Kinsale Boatyard +353 21477 4774
KPB Beaucette 07781 152581
Lake Yard Poole 01202 674531
Lallow, C Isle of Wight 01983 292112
Latham's Boatyard
Poole 01202 748029

Laxey Towing
Douglas, Isle of Man 07624 493592

Leonard Marine, Peter
Newhaven 01273 515987

Lincombe Marine
Salcombe 01548 843580

Lomax Boatbuilders
Cliffony +353 71 66124

Lymington Yt Haven 01590 677071

MacDougalls Marine Services
Isle of Mull 01681 700294

Macduff Shipyard Ltd
Macduff 01261 832234

Madog Boatyard
Porthmadog 01766 514205/513435

Mainbrayce Marine
Alderney 01481 822772

Malakoff and Moore
Lerwick 01595 695544

Mallaig Boat Building and Engineering
Mallaig 01687 462304

Maramarine
Helensburgh 01436 810971

Marindus Engineering
Kilmore Quay +353 53 29794

Mariners Farm Boatyard
Gillingham 01634 233179

McGruar and Co Ltd
Helensburgh 01436 831313

Mevagh Boatyard
Mulroy Bay +353 74 915 4470

Mill, Dan, Galway +353 86 337 9304

Mitchell's Boatyard
Poole 01202 747857

Mooney Boats
Killybegs +353 73 31152/31388

Moore & Son, J
St Austell 01726 842964

Morrison, A Killyleagh 028 44828215

Moss (Boatbuilders), David
Thornton-Cleveleys 01253 893830

Mustang Marine
Milford Haven 01646 696320

New Horizons Rhu 01436 821555

Noble and Sons, Alexander
Girvan 01465 712223

North Pier (Oban)
Oban 01631 562892

North Wales Boat Centre
Conwy 01492 580740

Northam Marine
Brightlingsea 01206 302003

Northshore Yacht Yard
Chichester 01243 512611

Oban Yachts and Marine Services
By Oban 01631 565333

Pearn and Co, Norman
Looe 01503 262244

Penrhos Marine
Aberdovey 01654 767478

Penzance Dry Dock and Engineering Co Ltd Penzance 01736 363838

Philip & Son Dartmouth 01803 833351

Phillips, HJ Rye 01797 223234

Pierce-Purcell
Galway +353 87 279 3821

Ponsharden Boatyard
Penryn 01326 372215

Powersail and Island Chandlers Ltd
East Cowes Marina 01983 299800

Priors Boatyard
Burnham-on-Crouch 01621 782160

R K Marine Ltd
Swanwick 01489 583572

Rat Island Sailboat Company (Yard) St Mary's 01720 423399

Rennison, Russell
Gosport 07734 688819

Retreat Boatyard Ltd
Exeter 01392 874720/875934

Rice and Cole Ltd
Burnham-on-Crouch 01621 782063

Richardson Boatbuilders, Ian Stromness 01856 850321

Richardsons Boatbuilders
Binfield 01983 821095

Riverside Yard
Shoreham Beach 01273 592456

River Yar Boatyard
Yarmouth, IoW 01983 761000

Robertsons Boatyard
Woodbridge 01394 382305

Rossbrin Boatyard
Schull +353 28 37352

Rossiter Yachts Ltd
Christchurch 01202 483250

Rossreagh Boatyard
Rathmullan +353 74 9150182

Rudders Boatyard & Moorings
Milford Haven 01646 600288

Ryan & Roberts Marine Services
Askeaton +353 61 392198

Rye Harbour Marina Rye
01797 227667

Rynn Engineering, Pat
Galway +353 91 562568

Salterns Boatyard Poole 01202 707391

Sandbanks Yacht Company
Poole 01202 611262

Scarborough Marine Engineering Ltd
Scarborough 01723 375199

Severn Valley Cruisers Ltd (Boatyard)
Stourport-on-Severn 01299 871165

Shepards Wharf Boatyard Ltd
Cowes 01983 297821

Shipshape
King's Lynn 01553 764058

Shotley Marina Ltd
Ipswich 01473 788982

Shotley Marine Services Ltd
Ipswich 01473 788913

Silvers Marina Ltd
Helensburgh 01436 831222

Skinners Boat Yard
Baltimore +353 28 20114

Smith, GB, & Sons
Rock 01208 862815

Smith & Gibbs
Eastbourne 07802 582009

Sparkes Boatyard
Hayling Island 023 92463572

Spencer Sailing Services, Jim
Brightlingsea 01206 302911

Standard House Boatyard
Wells-next-the-Sea 01328 710593

Storrar Marine Store
Newcastle upon Tyne 0191 266 1037

Strand Shipyard Rye 01797 222070

Surry Boatyard
Shoreham-by-Sea 01273 461491

The Shipyard
Littlehampton 01903 713327

Titchmarsh Marina
Walton-on-the-Naze 01255 672185

Tollesbury Marina
Tollesbury 01621 869202

T J Rigging Conwy 07780 972411

Toms and Son Ltd, C
Fowey 01726 870232

Tony's Marine Service
Coleraine 028 7035 6422

Torquay Marina
Torquay 01803 200210

Trinity Marine & Holyhead Marina
Holyhead 01407 763855

Trouts Boatyard (River Exe)
Topsham 01392 873044

Upson and Co, RF
Aldeburgh 01728 453047

Versatility Workboats
Rye 01797 224422

Weir Quay Boatyard
Bere Alston 01822 840474

West Solent Boatbuilders
Lymington 01590 642080

WicorMarine Fareham 01329 237112

Woodrolfe Boatyard
Maldon 01621 869202

Yarmouth Marine Services
Yarmouth, IoW 01983 760521

BOAT DELIVERIES & STORAGE

ABC Marine
Hayling Island 023 9246 1968

Abersoch Boatyard Services Ltd
Pwllheli 01758 713900

Ambrisbeg Ltd
Port Bannatyne 01700 502719

Arisaig Marine
Inverness-shire 01687 450224

Bedwell and Co
Walton-on-the-Naze 01255 675873

Berthon Boat Company
Lymington 01590 673312

Boat Shifters
07733 344018/01326 210548

C & J Marine Services
Newcastle upon Tyne 0191 295 0072

Caley Marine
Inverness 01463 233437

Carrick Marine Projects
Co Antrim 02893 355884

Challenger Marine
Penryn 01326 377222

Coates Marine Ltd
Whitby 01947 604486

Convoi Exceptionnel Ltd
Hamble 023 8045 3045

Creekside Boatyard (Old Mill Creek) Dartmouth 01803 832649

Crinan Boatyard Ltd
Crinan 01546 830232

Dale Sailing Co Ltd
Neyland 01646 603110

Dart Marina Ltd
Dartmouth 01803 833351

Dartside Quay
Brixham 01803 845445

Dauntless Boatyard Ltd
Canvey Island 01268 793782

Debbage Yachting
Ipswich 01473 601169

Douglas Marine
Preston 01772 812462

East & Co, Robin
Kingsbridge 01548 531257

East Coast Offshore Yachting
01480 861381

Emsworth Yacht Harbour
Emsworth 01243 377727

Exeter Ship Canal 01392 274306

Exmouth Marina 01395 269314

Firmhelm Ltd Pwllheli 01758 612244

Forrest Marine Ltd
Exeter 08452 308335

Fowey Boatyard
Fowey 01726 832194

Freshwater Boatyard
Truro 01326 270443

Hafan Pwllheli Pwllheli 01758 701219

Houghton Boat Transport
Tewkesbury 07831 486710

Gweek Quay Boatyard
Helston 01326 221657

Iron Wharf Boatyard
Faversham 01795 536296

Jalsea Marine Services Ltd
Northwich 01606 77870

KG McColl Oban 01852 200248

Latham's Boatyard
Poole 01202 748029

Lincombe Boat Yard
Salcombe 01548 843580

Marine Blast Holy Loch 01369 705394

Marine Resource Centre Ltd
Oban 01631 720291

Marine & General Engineers
Guernsey 01481 245808

Milford Marina
Milford Haven 01646 696312/3

Moonfleet Sailing
Poole 01202 682269

Murphy Marine Services
Cahersiveen +353 66 9476365

Southerly Chichester 01243 512611

Pasco's Boatyard Truro 01326 270269

Pearn and Co, Norman
Looe 01503 262244

Performance Yachting
Plymouth 01752 565023

Peters & May Ltd
Southampton 023 8048 0480

Ponsharden Boatyard
Penryn 01326 372215

Portsmouth Marine Engineering
Fareham 01329 232854

Priors Boatyard
Burnham-on-Crouch 01621 782160

Reeder School of Seamanship, Mike
Lymington 01590 674560

Rossiter Yachts
Christchurch 01202 483250

Sealand Boat Deliveries Ltd
Liverpool 01254 705225

Shearwater Sailing
Southampton 01962 775213

Shepards Wharf Boatyard Cowes Harbour Commission
Cowes 01983 297821

Silvers Marina Ltd
Helensburgh 01436 831222

Southcoasting Navigators
Devon 01626 335626

Waterfront Marine
Bangor 01248 352513

West Country Boat Transport
01566 785651

WicorMarine
Fareham 01329 237112

Winters Marine Ltd
Salcombe 01548 843580

Wolff, David 07659 550131**Yacht Solutions Ltd**
Portsmouth 023 9275 5155

Yarmouth Marine Service
Yarmouth, IoW 01983 760521

Youngboats
Faversham 01795 536176

BOOKS, CHARTS & PUBLISHERS

Adlard Coles Nautical
London 020 7631 5600

Brown Son & Ferguson Ltd
Glasgow 0141 429 1234

Cooke & Son Ltd, B
Hull 01482 223454

Dubois Phillips & McCallum Ltd
Liverpool 0151 236 2776

Imray, Laurie, Norie & Wilson
Huntingdon 01480 462114

Kelvin Hughes
Southampton 023 8063 4911

Lilley & Gillie Ltd, John 0191 257 2217

Marine Chart Services
Wellingborough 01933 441629

Price & Co Ltd, WF
Bristol 0117 929 2229

QPC
Fareham 01329 287880

Stanford Charts
Bristol 0117 929 9966

Stanford Charts
London 020 7836 1321

Stanford Charts 0845 880 3730
Manchester 0870 890 3730

Wiley Nautical
Chichester 01243 779777

BOW THRUSTERS

ARS Anglian Diesels Ltd
Wakefield 01924 332492

Buckler's Hard Boat Builders Ltd
Beaulieu 01590 616214

JS Mouldings International
Bursledon 023 8063 4400

BREAKDOWN

BJ Marine Ltd
Bangor, Ireland 028 9127 1434

Seastart
National 0800 885500

CHANDLERS

ABC Powermarine
Beaumaris 01248 811413

Admiral Marine Supplies
Bootle 01469 575909

Allgadgets.co.uk
Exmouth 01395 227727

Alpine Room & Yacht Equipment
Chelmsford 01245 223563

Aquatogs Cowes 01983 295071

Arbroath Fishermen's Association
Arbroath 01241 873132

Ardfern Yacht Centre Ltd
Argyll 01852 500247

Ardoran Marine
Oban 01631 566123

Arthurs Chandlery
Gosport 023 9252 6522

Arun Canvas and Rigging Ltd
Littlehampton 01903 732561

Aruncraft Chandlers
Littlehampton 01903 713327

ASAP Supplies – Equipment & Spares Worldwide
Beccles 0845 1300870

Auto Marine
Southsea 02392 825601

Bayside Marine
Brixham 01803 856771

Bedwell and Co
Walton on the Naze 01255 675873

BJ Marine Ltd Bangor 028 9127 1434

Bluecastle Chandlers
Portland 01305 822298

Blue Water Marine Ltd
Pwllheli 01758 614600

Boatacs
Westcliffe on Sea 01702 475057

Boathouse, The
Penryn 01326 374177

Booley Galway +353 91 562869

Boston Marina 01205 364420

Bosun's Locker, The
Falmouth 01326 312212

Bosun's Locker, The
Ramsgate 01843 597158

Bosuns Locker, The
South Queensferry 0131 331 3875/4496

B+ St Peter Port 01481 726071

Bridger Marine, John
Exeter 01392 250970

Bristol Boat Ltd Bristol 01225 872032

Brixham Yacht Supplies Ltd
Brixham 01803 882290

Brunel Chandlery Ltd
Neyland 01646 601667

Buccaneer Ltd
Macduff 01261 835199

Bucklers Hard Boat Builders
Beaulieu 01590 616214

Burghead Boat Centre
Findhorn 01309 690099

Bussell & Co, WL
Weymouth 01305 785633

Buzzard Marine
Yarmouth 01983 760707

C & M Marine
Bridlington 01262 672212

Cabin Yacht Stores
Rochester 01634 718020

Caley Marina Inverness 01463 236539

Cambrian Boat Centre
Swansea 01792 655925

Cantell & Son Ltd
Newhaven 01273 514118

Captain Watts Plymouth 01752 927067

Carne (Sales) Ltd, David
Penryn 01326 374177

Carrickcraft
Malahide +353 1 845 5438

Caters Carrick Ltd
Carrickfergus 028 93351919

CH Marine (Cork) +353 21 4315700

CH Marine Skibbereen +353 28 23190

Charity & Taylor Ltd
Lowestoft 01502 581529

Chertsey Marine Ltd
Penton Hook Marina 01932 565195

Chicks Marine Ltd
Guernsey 01481 740771

Christchurch Boat Shop
Christchurch 01202 482751

Clancy Hardware
Kilrush +35 65 905 1085

Clapson & Son (Shipbuilders) Ltd South Ferriby Marina 01652 635620

Clarke, Albert, Marine
Newtownards 028 9187 2325

Clyde Chandlers
Ardrossan 01294 607077

CMC Campbeltown 01586 551441

Coastal Marine Boatbuilders Ltd
(Dunbar) Eyemouth 01890 750328

Coates Marine Ltd
Whitby 01947 604486

Collins Marine
St Helier 01534 732415

Compass Marine
Lancing 01903 761773

Cosalt International Ltd
Aberdeen 01224 588327

Cosalt International Ltd
Southampton 023 8063 2824

Cotter, Kieran
Baltimore +353 28 20106

Cox Yacht Charter Ltd, Nick
Lymington 01590 673489

C Q Chandlers Ltd Poole 01202 682095

Crinan Boats Ltd
Lochgilphead 01546 830232

CTC Marine & Leisure
Middlesbrough 01642 372600

Dale Sailing Co Ltd
Milford Haven 01646 603110

Danson Marine
Sidcup 0208 304 5678

Dartmouth Chandlery
Dartmouth 01803 839292

Dartside Quay
Brixham 01803 845445

Dauntless Boatyard Ltd
Canvey Island 01268 793782

Davis's Yacht Chandler
Littlehampton 01903 722778

Denney & Son, EL
Redcar 01642 483507

Deva Marine Conwy 01492 572777

Dickie & Sons Ltd, AM
Bangor 01248 363400

Dickie & Sons Ltd, AM
Pwllheli 01758 701828

Dinghy Supplies Ltd/Sutton Marine Ltd
Sutton +353 1 832 2312

Diverse Yacht Services
Hamble 023 80453399

Dixon Chandlery, Peter
Exmouth 01395 273248

Doling & Son, GW
Barrow In Furness 01229 823708

Dovey Marine
Aberdovey 01654 767581

Down Marine Co Ltd
Belfast 028 9048 0247

Douglas Marine Preston 01772 812462

Dubois Phillips & McCallum Ltd
Liverpool 0151 236 2776

Duncan Ltd, JS Wick 01955 602689

Duncan Yacht Chandlers
Ely 01353 663095

East Anglian Sea School
Ipswich 01473 659992

Eccles Marine Co
Middlesbrough 01642 372600

Ely Boat Chandlers
Hayling Island 023 9246 1968

Emsworth Chandlery
Emsworth 01243 375500

Force 4 Chandlery
Stroud 0845 1300710

Exe Leisure Exeter 01392 879055

Express Marine Services
Chichester 01243 773788

Fairways Chandlery
Burnham-on-Crouch 01621 782659

Fairweather Marine
Fareham 01329 283500

Fal Chandlers
Falmouth Marina 01326 212411

Ferrypoint Boat Co
Youghal +353 24 94232

Findhorn Marina & Boatyard
Findhorn 01309 690099

Firmhelm Ltd P
wllheli 01758 612244

Fisherman's Mutual Asssociation (Eyemouth) Ltd
Eyemouth 01890 750373

Floetree Ltd (Loch Lomond Marina)
Balloch 01389 752069

Force 4 (Deacons)
Bursledon 023 8040 2182

Force 4 Chichester 01243 773788

Force 4 Chandlery
Mail order 0845 1300710

Force 4 Chandlery
Plymouth 01752 252489

Force 4 (Hamble Point)
Southampton 023 80455 058

Force 4 (Mercury)
Southampton 023 8045 4849

Force 4 (Port Hamble)
Southampton 023 8045 4858

Force 4 (Shamrock)
Southampton 023 8063 2725

Force 4 (Swanwick)
Swanwick 01489 881825

Freeport Marine
Jersey 01534 888100

French Marine Motors Ltd
Brightlingsea 01206 302133

Furneaux Riddall & Co Ltd
Portsmouth 023 9266 8621

Gael Force Glasgow 0141 941 1211

Gael Force Stornoway 01851 705540

Gallichan Marine Ltd
Jersey 01534 746387

Galway Maritime
Galway +353 91 566568

GB Attfield & Company
Dursley 01453 547185

Gibbons Ship Chandlers Ltd
Sunderland 0191 567 2101

Goodwick Marine
Fishguard 01348 873955

Gorleston Marine Ltd
Great Yarmouth 01493 661883

GP Barnes Ltd
Shoreham 01273 591705/596680

Great Outdoors
Clarenbridge, Galway +353 87 2793821

Green Marine, Jimmy
Fore St Beer 01297 20744

Grimsby Rigging Services Ltd
Grimsby 01472 362758

Gunn Navigation Services, Thomas
Aberdeen 01224 595045

Hale Marine, Ron
Portsmouth 023 92732985

Harbour Marine Services Ltd (HMS)
Southwold 01502 724721

Hardware & Marine Supplies
Wexford +353 53 29791

Hartlepool Marine Supplies
Hartlepool 01429 862932

Harwoods Yarmouth 01983 760258

Hawkins Marine Shipstores, John
Rochester 01634 840812

Hayles, Harold
Yarmouth 01983 760373

Herm Seaway Marine Ltd
St Peter Port 01481 726829

Highway Marine
Sandwich 01304 613925

Hoare Ltd, Bob, Poole 01202 736704

Hodges, T Coleraine 028 7035 6422

Iron Stores Marine
St Helier 01534 877755

Isles of Scilly Steamship Co
St Mary's 01720 422710

Jackson Yacht Services
Jersey 01534 743819

Jamison and Green Ltd
Belfast 028 9032 2444

Jeckells and Son Ltd
Lowestoft 01502 565007

JF Marine Chandlery
Rhu 01436 820584

JNW Services Aberdeen 01224 594050

JNW Services Peterhead 01779 477346

JSB Ltd
Tarbert, Loch Fyne 01880 820180

Johnston Brothers
Mallaig 01687 462215

Johnstons Marine Stores
Lamlash 01770 600333

Kearon Ltd, George
Arklow +353 402 32319

Kelvin Hughes Ltd
Southampton 023 80634911

Kildale Marine Hull 01482 227464

Kingfisher Marine
Weymouth 01305 766595

Kings Lock Chandlery
Middlewich 01606 737564

Kip Chandlery Inverkip
Greenock 01475 521485

Kirkcudbright Scallop Gear Ltd
Kirkcudbright 01557 330399

Kyle Chandlers Troon 01292 311880

Landon Marine, Reg
Truro 01872 272668

Largs Chandlers Largs 01475 686026

Lencraft Boats Ltd
Dungarvan +353 58 68220

Lincoln Marina Lincoln 01522 526896

Looe Chandlery
West Looe 01503 264355

Lynch Ltd, PA Morpeth 01670 512291

Mackay Boatbuilders (Arbroath) Ltd
Aberdeen 01241 872879

Mackay Marine Services
Aberdeen 01224 575772

Mailspeed Marine
Crawley 01273837823

Mailspeed Marine
Essex Marina 01342 710618

Mailspeed Marine
Warrington 01342 710618

Mainbrayce Chandlers
Braye, Alderney 01481 822772

Manx Marine Ltd
Douglas 01624 674842

Marine & Leisure Europe Ltd
Plymouth 01752 268826

Marine MegaStore
Hamble 023 8045 4400

Marine MegaStore
Morpeth 01670 516151

Marine Parts Direct
Swords, Co Dublin +353 1 807 5144

Marine Scene Bridgend 01656 671822

Marine Scene
Cardiff 029 2070 5780

Marine Services
Jersey 01534 626930

Marine Store Wyatts
West Mersea 01206 384745

Marine Store
Maldon 01621 854280

Marine Store
Titchmarsh Marina 01255 679028

Marine Store
Walton on the Naze 01255 679028

Marine Superstore Port Solent Chandlery
Portsmouth 023 9221 9843

MarineCo
Torpoint 01752 816005

Maryport Harbour and Marina
Maryport 01900 814431

Matthews Ltd, D
Cork +353 214 277633

McClean
Greenock 01475 728234

McCready Sailboats Ltd
Holywood 028 9042 1821

Moore, Kevin Cowes 01983 289699

Moore & Son, J
Mevagissey 01726 842964

Morgan & Sons Marine, LH
Brightlingsea 01206 302003

Mount Batten Boathouse
Plymouth 01752 482666

Murphy Marine Services
Cahersiveen +353 66 9476365

Murphy, Nicholas
Dunmore East +353 51 383259

MUT Wick 07753 350143

Mylor Chandlery & Rigging
Falmouth 01326 375482

Nautical World Bangor 028 91460330

New World Yacht Care
Helensburgh 01436 820586

Newhaven Chandlery
Newhaven 01273 612612

Nifpo Ardglass 028 4484 2144

Norfolk Marine
Great Yarmouth 01692 670272

Norfolk Marine Chandlery Shop
Norwich 01603 783150

Ocean Leisure Ltd
London 020 7930 5050

One Stop Chandlery
Maldon 01621 853558

O'Sullivans Marine Ltd
Tralee +353 66 7129635

Partington Marine Ltd, William
Pwllheli 01758 612808

Pascall Atkey & Sons Ltd
Isle of Wight 01983 292381

Pennine Marine Ltd
Skipton 01756 792335

Penrhos Marine Aberdovey 01654 767478
Penzance Marine Services Penzance 01736 361081
Perry Marine, Rob Axminster 01297 631314
Pepe Boatyard Hayling Island 023 9246 1968
Performance Yachting & Chandlery Plymouth 01752 565023
Pinnell & Bax Northampton 01604 592808
Piplers of Poole Poole 01202 673056
Pirate's Cave, The Rochester 01634 295233
Powersail Island Chandlers Ltd East Cowes Marina 01983 299800
Preston Marine Services Ltd Preston 01772 733595
Price & Co Ltd, WF Bristol 0117 929 2229
PSM Ltd Alderney 07781 106635
Purcell Marine Clarenbridge +353 87 279 3821
Purple Sails & Marine Walsall 08456 435510
Quay West Marine Poole 01202 732445
Quayside Marine Salcombe 01548 844300
R&A Fabrication Kirkcudbright 01557 330399
Racecourse Yacht Basin (Windsor) Ltd Windsor 01753 851501
Rat Rigs Water Sports Cardiff 029 2062 1309
Reliance Marine Wirral 0151 625 5219
Rigmarine Padstow 01841 532657
Riversway Marine Preston 0844 879 4901
RHP Marine Cowes 01983 290421
RNS Marine Northam 01237 474167
Sail Loft Bideford 01271 860001
Sailaway St Anthony 01326 231357
Salcombe Boatstore Salcombe 01548 843708
Salterns Chandlery Poole 01202 701556
Shipmate Salcombe 01548 844555
Sandrock Marine Rye 01797 222679
Schull Watersports Centre Schull +353 28 28554
Sea & Shore Ship Chandler Dundee 01382 450666
Sea Cruisers of Rye Rye 01797 222070
Sea Span Edinburgh 0131 552 2224
Sea Teach Ltd Emsworth 01243 375774
Seafare Tobermory 01688 302277
Seahog Boats Preston 01772 633016
Seamark-Nunn & Co Felixstowe 01394 451000
Seaquest Marine Ltd St Peter Port 01481 721773
Seaware Ltd Penryn 01326 377948
Seaway Marine Macduff 01261 832877
Sharp & Enright Dover 01304 206295
Shearwater Engineering Services Ltd Dunoon 01369 706666
Shipshape Marine King's Lynn 01553 764058
Ship Shape Ramsgate 01843 597000
Shorewater Sports Chichester 01243 672315
Simpson Marine Ltd Newhaven 01273 612612
Simpson Marine Ltd, WA Dundee 01382 566670
Sketrick Marine Centre Killinchy 028 9754 1400
Smith AM (Marine) Ltd London 020 8529 6988
Solent Marine Chandlery Ltd Gosport 023 9258 4622
South Coast Marine Christchurch 01202 482695
South Pier Shipyard St Helier 01534 711000
Southampton Yacht Services Ltd Southampton 023 803 35266
Sparkes Chandlery Hayling Island 02392 463572
S Roberts Marine Ltd Liverpool 0151 707 8300
SSL Marine Eastbourne 01323 47900
Standard House Chandlery Wells-next-the-Sea 01328 710593
Stornoway Fishermen's Co-op Stornoway 01851 702563
Sunset Marine & Watersports Sligo +353 71 9162792
Sussex Marine St Leonards on Sea 01424 425882
Sussex Yachts Ltd Shoreham 01273 605482
Sussex Marine Centre Shoreham 01273 454737
Sutton Marine (Dublin) Sutton +353 1 832 2312
SW Nets Newlyn 01736 360254
Tarbert Ltd, JSB Tarbert 01880 820180
TCS Chandlery Essex Marina 01702 258094
TCS Chandlery Grays 01375 374702
TCS Chandlery Southend 01702 444423
Thulecraft Ltd Lerwick 01595 693192
Tony's Marine Services Coleraine 07866 690436
Torbay Boating Centre Paignton 01803 558760
Torquay Chandlers Torquay 01803 211854
Trafalgar Yacht Services Fareham 01329 822445
Trident UK N Shields 0191 490 1736
Union Chandlery Cork +353 21 4554334
Uphill Boat Services Weston-Super-Mare 01934 418617
Upper Deck Marine and Outriggers Fowey 01726 832287
V Ships (Isle of Man) Douglas 01624 688886
V F Marine Rhu 01436 820584
Viking Marine Ltd Dun Laoghaire +353 1 280 6654
Walker Boat Sales Deganwy 01492 555706
Waterfront Marine Bangor 01248 352513
Wayne Maddox Marine Margate 01843 297157
Western Marine Dalkey +353 1280 0321
Wetworks, The Burnham-on-Crouch 01621 786413
Whitstable Marine Whitstable 01227 274168
Williams Ltd, TJ Cardiff 029 20 487676
Windjammer Marine Milford Marina 01646 699070
Yacht & Boat Chandlery Faversham 01795 531777
Yacht Chandlers Conwy 01492 572777
Yacht Equipment Chelmsford 01245 223563
Yachtmail Ltd Lymington 01590 672784
Yachtshop Conwy 01492 338505
Yachtshop Holyhead 01407 760031
You Boat Chandlery Gosport 02392 522226

CHART AGENTS

Brown Son & Ferguson Ltd Glasgow 0141 429 1234
Chattan Security Ltd Edinburgh 0131 554 7527
Cooke & Son Ltd, B Hull 01482 223454
Dubois Phillips & McCallum Ltd Liverpool 0151 236 2776
Imray Laurie Norie and Wilson Ltd Huntingdon 01480 462114
Kelvin Hughes Southampton 023 8063 4911

Lilley & Gillie Ltd, John
North Shields 0191 257 2217

Marine Chart Services
Wellingborough 01933 441629

Sea Chest Nautical Bookshop
Plymouth 01752 222012

Seath Instruments (1992) Ltd
Lowestoft 01502 573811

Small Craft Deliveries
Woodbridge 01394 382655

Smith (Marine) Ltd, AM
London 020 8529 6988

South Bank Marine Charts Ltd Grimsby
01472 361137

Stanford Charts Bristol 0117 929 9966

Stanford Charts
London 020 7836 1321

Todd Chart Agency Ltd
County Down 028 9146 6640

UK Hydrographics Office
Taunton 01823 337900

Warsash Nautical Bookshop
Warsash 01489 572384

CLOTHING

Absolute
Gorleston on Sea 01493 442259

Aquatogs Cowes 01983 245892

Crew Clothing
London 020 8875 2300

Crewsaver
Gosport 01329 820000

Douglas Gill
Nottingham 0115 9460844

Fat Face fatface.com

Gul International Ltd
Bodmin 01208 262400

Guy Cotten UK Ltd
Liskeard 01579 347115

Harwoods
Yarmouth 01983 760258

Helly Hansen
Nottingham 0115 979 5997

Henri Lloyd
Manchester 0161 799 1212

Joules 0845 6066871

Mad Cowes Clothing Co
Cowes 0845 456 5158

Matthews Ltd, D
Cork +353 214 277633

Mountain & Marine
Poynton 01625 859863

Musto Ltd Laindon 01268 491555

Purple Sails & Marine
Walsall 0845 6435510

Quba Sails
Lymington 01590 689362

Quba Sails Salcombe 01548 844599

Yacht Parts Plymouth 01752 252489

CODE OF PRACTICE EXAMINERS

Booth Marine Surveys, Graham
Birchington-on-Sea 01843 843793

Cannell & Associates, David M
Wivenhoe 01206 823337

COMPUTERS & SOFTWARE

Dolphin Maritime Software
White Cross 01524 841946

Forum Software Ltd
Nr Haverfordwest 01646 636363

Kelvin Hughes Ltd
Southampton 023 8063 4911

Memory-Map
Aldermaston 0844 8110950

PC Maritime Plymouth 01752 254205

DECK EQUIPMENT

Aries Van Gear Spares
Penryn 01326 377467

Ronstan Gosport 023 9252 5377

Harken UK Lymington 01590 689122

IMP Royston 01763 241300

Kearon Ltd George +353 402 32319

Pro-Boat Ltd
Burnham-on-Crouch 01621 785455

Ryland, Kenneth
Stanton 01386 584270

Timage & Co Ltd
Braintree 01376 343087

DIESEL MARINE/ FUEL ADDITIVES

Corralls Poole 01202 674551

Cotters Marine & General Supplies
Baltimore +353 28 20106

Expresslube
Henfield 01444 254115

Gorey Marine Fuel Supplies
Gorey 07797 742384

Hammond Motorboats
Dover 01304 206809

Iron Wharf Boatyard
Faversham 01795 536296

Lallow, Clare Cowes 01983 760707

Marine Support & Towage
Cowes 01983 200716/07860 297633

McNair, D
Campbeltown 01586 552020

Quayside Fuel
Weymouth 07747 182181

Rossiter Yachts
Christchurch 01202 483250

Sleeman & Hawken
Shaldon 01626 778266

DIVERS

Abco Divers Belfast 028 90610492

Andark Diving
Burseldon 01489 581755

Argonaut Marine
Aberdeen 01224 706526

Baltimore Diving and Watersports Centre West Cork +353 28 20300

C & C Marine Services
Largs 01475 687180

Cardiff Commercial Boat Operators Ltd
Cardiff 029 2037 7872

Clyde Diving Centre
Inverkip 01475 521281

Divetech UK King's Lynn 01485 572323

Diving & Marine Engineering
Barry 01446 721553

Donnelly, R
South Shields 07973 119455

DV Diving 028 9146 4671

Falmouth Divers Ltd
Penryn 01326 374736

Fathoms Ltd Wick 01955 605956

Felixarc Marine Ltd
Lowestoft 01502 509215

Grampian Diving Services
New Deer 01771 644206

Higgins, Noel +353 872027650

Hudson, Dave
Trearddur Bay 01407 860628

Hunt, Kevin Tralee +353 6671 25979

Kaymac Diving Services
Swansea 08431 165523

Keller, Hilary Buncrana +353 77 62146

Kilkee Diving Centre
Kilkee +353 6590 56707

Leask Marine Kirkwall 01856 874725

Looe Divers Hannafore 01503 262727

MacDonald, D Nairn 01667 455661

Medway Diving Contractors Ltd
Gillingham 01634 851902

MMC Diving Services
Lake, Isle of Wight 07966 579965

Mojo Maritime
Penzance 01736 762771

Murray, Alex
Stornoway 01851 704978

New Dawn Dive Centre
Lymington 01590 675656

Northern Divers (Engineering) Ltd
Hull 01482 227276

Offshore Marine Services Ltd
Bembridge 01983 873125

Parkinson (Sinbad Marine Services), J
Killybegs +353 73 31417

Port of London Authority
Gravesend 01474 560311

Purcell, D – Crouch Sailing School
Burnham 01621 784140/0585 33

Salvesen UK Ltd
Liverpool 0151 933 6038

Sea-Lift Diving Dover 01304 829956

Southern Cylinder Services
Fareham 01329 221125

Sub Aqua Services
North Ormesby 01642 230209

Teign Diving Centre
Teignmouth 01626 773965

Tuskar Rock Marine
Rosslare +353 53 33376

Underwater Services
Dyffryn Arbwy 01341 247702

Wilson Alan c/o Portrush Yacht Club
Portrush 028 2076 2225

Woolford, William
Bridlington 01262 671710

ELECTRICAL AND ELECTRONIC ENGINEERS

AAS Marine
Aberystwyth 01970 631090

Allworth Riverside Services, Adrian
Chelsea Harbour Marina
07831 574774

ASL Auto Services
Boston 01205 761560

Auto Marine Electrics
Holy Loch 01369 701555

Baker, Keith Brentford 07792 937790

Belson Design Ltd, Nick
Southampton 077 6835 1330

Biggs, John Weymouth Marina,
Weymouth 01305 778445

BJ Marine Ltd Bangor 028 9127 1434

BM Electrical
Kirkcudbright 07584 657192

Boat Electrics Troon 01292 315355

Buccaneer Ltd Macduff 01261 835199

Calibra Marine
Dartmouth 01803 833094

Campbell & McHardy Lossiemouth
Marina, Lossiemouth 01343 812137

CES Sandown Sparkes Marina,
Hayling Island 023 9246 6005

Colin Coady Marine
Malahide +353 87 265 6496

Contact Electrical
Arbroath 01241 874528

DDZ Marine Ardossan 01294 607077

Dobson, Chris
Whitehaven 07986 086641

EC Leisure Craft
Essex Marina 01702 568482

Energy Solutions
Rochester 01634 290772

Enterprise Marine Electronic & Technical Services Ltd
Aberdeen 01224 593281

Eurotek Marine
Eastbourne 01323 479144

Evans, Lyndon
Brentford 07795 218704

Floetree Ltd (Loch Lomond Marina)
Balloch 01389 752069

Hamble Marine
Hamble 02380 001088

HNP Engineers (Lerwick) Ltd
Lerwick 01595 692493

Jackson Yacht Services
Jersey 01534 743819

Jedynak, A Salcombe 01548 843321

Kippford Slipway Ltd
Dalbeattie 01556 620249

Lynch Ltd, PA Morpeth 01670 512291

Lynch, John Tralee +353 87 992 3102

Mackay Boatbuilders (Arbroath) Ltd
Aberdeen 01241 872879

Marine, AW Gosport 023 9250 1207

Marine Electrical Repair Service
London 020 7228 1336

Maxfield Electrical
Doncaster 07976 825349

MB Marine Troon 01292 311944

McMillan, Peter
Kilrush +35 86 8388617

MES Falmouth Marina,
Falmouth 01326 378497

Mount Batten Boathouse
Plymouth 01752 482666

New World Yacht Care
Rhu 01436 820586

Neyland Marine Services Ltd
Milford Haven 01646 600358

Powell, Martin Shamrock Quay,
Southampton 023 8033 2123

PR Systems
Plymouth 01752 936145

Radio & Electronic Services Beaucette Marina,
Guernsey 01481 728837

RHP Marine Cowes 01983 290421

Rothwell, Chris
Torquay Marina 01803 850960

Ruddy Marine
Galway +353 87 742 7439

Rutherford, Jeff Largs 01475 568026

SM International
Plymouth 01752 662129

Sussex Fishing Services
Rye 01797 223895

Tony's Marine Services
Coleraine 07866 690436

Ultra Marine Systems
Mayflower International Marina,
Plymouth 07989 941020

Upham, Roger
Chichester 01243 514511

Volspec Ipswich 01473 780144

Weyland Marine Services
Milford Haven 01646 600358

ELECTRONIC DEVICES AND EQUIPMENT

Anchorwatch UK
Edinburgh 0131 447 5057

Aquascan International Ltd
Newport 01633 841117

Atlantis Marine Power Ltd
Plymouth 01752 208810

Autosound Marine
Bradford 01274 688990

B&G Romsey 01794 518448

Brookes & Gatehouse
Romsey 01794 518448

Boat Electrics & Electronics Ltd
Troon 01292 315355

Cactus Navigation & Communication
London 020 7833 3435

CDL Aberdeen 01224 706655

Charity & Taylor Ltd
Lowestoft 01502 581529

Diverse Yacht Services
Hamble 023 8045 3399

Dyfed Electronics Ltd
Milford Haven 01646 694572

Echopilot Marine Electronics Ltd
Ringwood 01425 476211

Enterprise Marine
Aberdeen 01224 593281

Euronav Ltd
Portsmouth 023 9298 8806

Exposure Lights
Pulborough 01798 83930

Furuno UK
Fraserburgh 01346 518300
Havant 023 9244 1000

Garmin (Europe) Ltd
Romsey 0870 850 1242

Golden Arrow Marine Ltd
Southampton 023 8071 0371

Greenham Regis Marine Electronics
Lymington 01590 671144

Greenham Regis Marine Electronics
Poole 01202 676363

Greenham Regis Marine Electronics
Southampton 023 8063 6555

ICS Electronics
Arundel 01903 731101

JG Technologies Ltd Weymouth 0845 458 9616
KM Electronics Lowestoft 01502 569079
Kongsberg Simrad Ltd Aberdeen 01224 226500
Kongsberg Simrad Ltd Wick 01955 603606
Landau UK Ltd Hamble 02380 454040
Enterprise Marine Aberdeen 01224 593281
Marathon Leisure Hayling Island 023 9263 7711
Marine Instruments Falmouth 01326 375483
MB Marine Troon 01292 311944
Microcustom Ltd Ipswich 01473 215777
Nasa Marine Instruments Stevenage 01438 354033
Navionics UK Plymouth 01752 204735
Ocean Leisure Ltd London 020 7930 5050
Plymouth Marine Electronics Plymouth 01752 227711
Radio & Electronic Services Ltd St Peter Port 01481 728837
Raymarine Ltd Portsmouth 02392 714700
Redfish Car Company Stockton-on-Tees 01642 633638
Robertson, MK Oban 01631 563836
Satcom Distribution Ltd Salisbury 01722 410800
Seaquest Marine Ltd Guernsey 01481 721773
Seatronics Aberdeen 01224 853100
Selex Communications Aberdeen 01224 890316
Selex Communications Bristol 0117 931 3550
Selex Communications Brixham 01803 882716
Selex Communications Fraserburgh 01346 518187
Selex Communications Glasgow 0141 882 6909
Selex Communications Hull 01482 326144
Selex Communications Kilkeel 028 4176 9009
Selex Communications Liverpool 01268 823400
Selex Communications Lowestoft 01502 572365
Selex Communications Newcastle upon Tyne 0191 265 0374
Selex Communications Newlyn 01736 361320
Selex Communications Penryn 01326 378031
Selex Communications Plymouth 01752 222878
Selex Communications Rosyth 01383 419606
Selex Communications Southampton 023 8051 1868
Silva Ltd Livingston 01506 419555
SM International Plymouth 01752 662129
Sperry Marine Ltd Peterhead 01779 473475
Stenmar Ltd Aberdeen 01224 827288
Transas Nautic Portsmouth 023 9267 4016
Veripos Precise Navigation Aberdeen 01224 965800
Wema (UK) Honiton 01404 881810
Wilson & Co Ltd, DB Glasgow 0141 647 0161
Woodsons of Aberdeen Ltd Aberdeen 01224 722884

ENGINES AND ACCESSORIES

Airylea Motors Aberdeen 01224 872891
Amble Boat Co Ltd Amble 01665 710267
Anchor Marine Products Benfleet 01268 566666
Aquafac Ltd Luton 01582 568700
Barrus Ltd, EP Bicester 01869 363636
British Polar Engines Ltd Glasgow 0141 445 2455
Bukh Diesel UK Ltd Poole 01202 668840
CJ Marine Mechanical Troon 01292 313400
Cleghorn Waring Ltd Letchworth 01462 480380
Cook's Diesel Service Ltd Faversham 01795 538553
Southern Shipwright (SSL) Brighton 01273 601779
Southern Shipwright (SSL) Eastbourne 01323 479000
Fender-Fix Maidstone 01622 751518
Fettes & Rankine Engineering Aberdeen 01224 573343
Fleetwood & Sons Ltd, Henry Lossiemouth 01343 813015
Gorleston Marine Ltd Great Yarmouth 01493 661883
Halyard Salisbury 01722 710922
Interseals (Guernsey) Ltd Guernsey 01481 246364
Kelpie Boats Pembroke Dock 01646 683661
Keypart Watford 01923 330570
Lancing Marine Brighton 01273 410025
Lencraft Boats Ltd Dungarvan +353 58 68220
Lewmar Ltd Havant 023 9247 1841
Liverpool Power Boats Bootle 0151 944 1163
Lynch Ltd, PA Morpeth 01670 512291
MacDonald & Co Ltd, JN Glasgow 0141 810 3400
Mariners Weigh Shaldon 01626 873698
MMS Ardrossan 01294 604831
Mooring Mate Ltd Bournemouth 01202 421199
Murphy Marine Services Cahersiveen +353 66 9476365
Newens Marine, Chas Putney 020 8788 4587
Ocean Safety Southampton 023 8072 0800
RK Marine Ltd Hamble 01489 583585
Swanwick 01489 583572
Sillette Sonic Ltd Sutton 020 8337 7543
Sowester Simpson-Lawrence Ltd Poole 01202 667700
Timage & Co Ltd Braintree 01376 343087
Thorne Boat Services Thorne 01405 814197
Vetus Den Ouden Ltd Totton 023 8045 4507
Western Marine Dublin +353 1 280 0321
Whitstable Marine Whitstable 01227 262525
Yates Marine, Martin Galgate 01524 751750
Ynys Marine Cardigan 01239 613179

FOUL-WEATHER GEAR

Aquatogs Cowes 01983 295071
Crew Clothing London 020 8875 2300
Century Finchampstead 0118 9731616
Crewsaver Gosport 01329 820000
Douglas Gill Nottingham 0115 946 0844

FBI Leeds 0113 270 7000
Gul International Ltd Bodmin 01208 262400
Helly Hansen Nottingham 0115 979 5997
Henri Lloyd Manchester 0161 799 1212
Musto Ltd Laindon 01268 491555
Pro Rainer Windsor 07752 903882

GENERAL MARINE EQUIPMENT & SPARES

Ampair Ringwood 01425 480780
Aries Vane Gear Spares Penryn 01326 377467
Arthurs Chandlery, R Gosport 023 9252 6522
Atlantis Marine Power Ltd Plymouth 01752 208810
Barden UK Ltd Fareham 01489 570770
Calibra Marine International Ltd Southampton 08702 400358
CH Marine (Cork) +353 21 4315700
Chris Hornsey (Chandlery) Ltd Southsea 023 9273 4728
Compass Marine (Dartmouth) Dartmouth 01803 835915
Cox Yacht Charter Ltd, Nick Lymington 01590 673489
CTC Marine & Leisure Middlesbrough 01642 372600
Exposure Lights Pulborough 01798 83930
Frederiksen Boat Fittings (UK) Ltd Gosport 023 9252 5377
Furneaux Riddall & Co Ltd Portsmouth 023 9266 8621
Hardware & Marine Supplies Co Wexford +353 (53) 29791
Index Marine Bournemouth 01202 470149
Kearon Ltd, George Arklow +353 402 32319
Marathon Leisure Hayling Island 023 9263 7711
Pro-Boat Ltd Burnham-on-Crouch 01621 785455
Pump International Ltd Cornwall 01209 831937
Quay West Marine Poole 01202 732445
Rogers, Angie Bristol 0117 973 8276
Ryland, Kenneth Stanton 01386 584270
Tiflex Liskeard 01579 320808
Vetus Boating Equipment Southampton 02380 454507
Whitstable Marine Whitstable 01227 262525
Yacht Parts Plymouth 01752 252489

HARBOUR MASTERS

Aberaeron 01545 571645
Aberdeen 01224 597000
Aberdovey 01654 767626
Aberystwyth 01970 611433
Alderney & Burhou 01481 822620
Amble 01665 710306
Anstruther 01333 312951
Appledore 01237 474569
Arbroath 01241 872166
Ardglass 028 4484 1291
Ardrossan Control Tower 01294 463972
Arinagour Piermaster 01879 230347
Arklow +353 402 32466
Baltimore +353 28 22145
Banff 01261 815544
Bantry Bay +353 27 53277
Barmouth 01341 280671
Barry 01446 732665
Beaucette 01481 245000
Beaulieu River 01590 616200
Belfast Lough 028 90 553012
Belfast River Manager 028 90 328507
Bembridge 01983 872828
Berwick-upon-Tweed 07931 730165
Bideford 01237 346131
Blyth (Port Ops) 01670 357025
Boston 01205 362328
Bridlington 01262 670148/9
Bridport 01308 423222
Brighton 01273 819919
Bristol 0117 926 4797
Brixham 01803 853321
Buckie 01542 831700
Bude 01288 353111
Burghead 01542 831700
Burnham-on-Crouch 01621 783602
Burnham-on-Sea 0300 303 7799
Burtonport +353 075 42155
Caernarfon 01286 672118
Caernarfon 07786 730865
Caledonian Canal Off. (Inverness) 01463 725500
Camber Berthing Offices – Portsmouth 023 92297395
Campbeltown 01586 552552 07825 732862
Canna 01687 310733
Cardiff 029 20400500
Carnlough Harbour 07703 606763
Castletown Bay 01624 823549
Charlestown 01726 67526
Chichester Harbour 01243 512301
Clovelly 01273 431549 07975 501380
Conwy 01492 596253
Cork +353 21 4273125
Corpach Canal Sea Lock 01397 772249
Courtmacsherry +353 8673 94299 +353 23 46311/46600
Coverack 01326 380679
Cowes 01983 293952
Crail 01333 450820
Craobh Haven 01852 502222
Crinan Canal Office 01546 603210
Cromarty Firth 01381 600479
Cromarty Harbour 01381 600493
Crookhaven +353 28 35319
Cullen 01542 831700
Dingle +353 66 9151629
Douglas 01624 686628
Dover 01304 240400 Ext 4520
Dublin +353 1 874871
Dun Laoghaire +353 1 280 1130/8074
Dunbar 07958 754858
Dundee 01382 224121
East Loch Tarbert 01859 502444
Eastbourne 01323 470099
Eigg Harbour 01687 482428
Elie 01333 330051
Estuary Control - Dumbarton 01389 726211
Exe 01392 274306
Exeter 01392 265791
Exmouth 01392 223265 07864 958658
Eyemouth 01890 750223 07885 742505
Felixstowe 07803 476621
Findochty 01542 831700
Fisherrow 0131 665 5900
Fishguard (Lower Harbour) 01348 874726
Fishguard 01348 404425

Fleetwood	01253 872323
Flotta	01856 701411
Folkestone	01303 715354
Fowey	01726 832471/2.
Fraserburgh	01346 515858
Galway Bay	+353 91 561874
Garlieston	01988 600274
Glasson Dock	07910 315606
Gorey Port Control	01534 447788
Gourdon	01569 762741
Great Yarmouth	01493 335501
Grimsby Dockmaster	01472 359181
Groomsport Bay	028 91 278040
Hamble River	01489 576387
Hayle	07500 993867
Helford River	01326 732544
Helmsdale	01431 821692
Holy Island	01289 389217
Holyhead	01407 763071
Hopeman	01542 831700
Howth	+353 1 832 2252
Ilfracombe	01271 862108
Inverness	01463 715715
Ipswich	01473 211771
Irvine	01294 487286
Johnshaven	01561 362262
Kettletoft Bay	01857 600227
Killybegs	+353 73 31032
Kilmore Quay	+353 53 912 9955
Kinlochbervie	01971 521235 07901 514350
Kinsale	+353 21 4772503
Kirkcudbright	01557 331135
Kirkwall	01856 872292
Langstone Harbour	023 9246 3419
Larne	02828 872100
Lerwick	01595 692991
Littlehampton	01903 721215
Liverpool	0151 949 6134/5
Loch Gairloch	01445 712140
Loch Inver	01571 844267 07958 734610
Looe	01503 262839 07918 728955
Lossiemouth (Marina)	07969 213513 07969 213521
Lough Foyle	028 7186 0555
Lowestoft	01502 572286
Lyme Regis	01297 442137
Lymington	01590 672014
Lyness	01856 791387
Macduff	01261 832236
Maryport	01900 814431
Menai Strait	01248 712312
Methil	01333 462725
Mevagissey	01726 843305
Milford Haven	01646 696100
Minehead (Mon-Fri)	01643 702566
Montrose	01674 672302
Mousehole	01736 731511
Mullion Cove	01326 240222
Nairn Harbour Office	01667 452453
Newhaven Harbour Admin	01273 612872/612926
Newlyn	01736 731897
Newquay	07737 387217
Newport Harbour Office	01983 823885
North Berwick	00776 467373
Oban	01631 562892
Padstow	01841 532239
Peel	01624 842338
Penrhyn Bangor	01248 352525
Penzance	01736 366113
Peterhead	01779 483630
Pierowall	01857 677216
Pittenweem	01333 312591
Plockton	01599 534589
Polperro	01503 272634 07966 528045
Poole	01202 440233
Port Ellen Harbour Association	01496 302458
Port Isaac	01208 880321 07855 429422
Port St Mary	01624 833205
Porth Dinllaen	01758 720276
Porthleven	01326 574207
Porthmadog	01766 512927
Portknockie	01542 831700
Portland	01305 824044
Portpatrick	01776 810355
Portree	01478 612926
Portrush	028 70822307
Portsmouth Harbour Commercial Docks	023 92297395
Portsmouth Harbour Control	023 92723694
Portsmouth Harbour	023 92723124
Preston	01772 726711
Pwllheli	01758 701219
Queenborough	01795 662051
Queens Gareloch/Rhu	01436 674321
Ramsey	01624 812245
Ramsgate	01843 572100
River Bann & Coleraine	028 7034 2012
River Blackwater	01621 856487
River Colne (Brightlingsea)	01206 302200
River Dart	01803 832337
River Deben	01473 736257
River Exe Dockmaster	01392 274306
River Humber	01482 327171
River Medway	01795 596593
River Orwell	01473 231010
River Roach	01621 783602
River Stour	01255 243000
River Tyne/N Shields	0191 257 2080
River Yealm	01752 872533
Rivers Alde & Ore	07528 092635
Rosslare Europort	+353 53 915 7921
Rothesay	01700 503842 07799 724225
Ryde	01983 613903
Salcombe	01548 843791
Sark	01481 832323
Scalloway	01595 880574
Schull	+353 27 28136
Scarborough	01723 373530
Scrabster	01847 892779
Seaham	07786 565205
Sharpness, Gloucester Harbour Trustees	01453 811913
Shoreham	01273 598100
Silloth	016973 31358
Sligo	+353 91 53819
Southampton	023 8033 9733
Southend-on-Sea	01702 611889
Southwold	01502 724712
St Helier	01534 447788
St Ives	07793 515460
St Margaret's Hope	01856 831454
St Mary's	01720 422768
St Michael's Mount	07870 400282
St Monans (part-time)	07930 869538
St Peter Port	01481 720229
Stonehaven	01569 762741
Stornoway	01851 702688
Strangford Lough	028 44 881637
Stranraer	07734 073421

Stromness 07810 465825
Stronsay 01857 616317
Sullom Voe 01806 242551
Sunderland 0191 567 2626
Swale 01795 561234
Swansea 01792 653787
Tayport Hbr Trust 01382 553799
Tees & Hartlepool Port Authority 01429 277205
Teignmouth 01626 773165
Tenby 01834 842717
Thames Estuary 01474 562200
Tobermory Moorings Officer 07917 832497
Torquay 01803 292429
Troon 01292 281687
Truro 01872 272130
Ullapool 01854 612091
Waldringfield 01394 736291
Walton-on-the-Naze 01255 851899
Watchet 01643 703704
Waterford +353 51 899801
Wells-next-the-Sea 01328 711646
West Bay (Bridport) 01308 423222
07870 240636
Wexford +353 53 912 2039
Weymouth 01305 206423
Whitby 01947 602354
Whitehaven 01946 692435
Whitehills 01261 861291
Whitstable 01227 274086
Wick 01955 602030
Wicklow +353 404 67455
Workington 01900 602301
Yarmouth 01983 760321
Youghal +353 24 92626
+353 86 780 0878

HARBOURS

Bristol Harbour 0117 903 1484
Clyde Marina – Ardrossan 01294 607077
Jersey Hbrs St Helier 01534 885588
Maryport Harbour and Marina Maryport 01900 818447/4431
Peterhead Bay Authority Peterhead 01779 474020
Sark Moorings – Channel Islands 01481 832260

INSURANCE/FINANCE

Admiral Marine Ltd Salisbury 01722 416106
Bishop Skinner Boat Insurance London 0800 7838057
Bluefin London 0800 074 5200
Castlemain Ltd St Peter Port 01481 721319
Craven Hodgson Associates Leeds 0113 243 8443
Giles Insurance Brokers Irvine 01294 315481
GJW Direct Liverpool 0151 473 8000
Haven Knox-Johnston West Malling 01732 223600
Lombard Southampton 023 8024 2171
Mercia Marine Malvern 01684 564457
Nautical Insurance Services Ltd Leigh-on-Sea 01702 470811
Navigators & General Brighton 01273 863400
Pantaenius UK Ltd Plymouth 01752 223656
Porthcawl Insurance Consultants Porthcawl 01656 784866
Saga Boat Insurance Folkestone 01303 771135
St Margarets Insurances London 020 8778 6161
Towergate Insurance Shrewsbury 0344 892 1987

LIFERAFTS & INFLATABLES

Adec Marine Ltd Croydon 020 8686 9717
Avon Inflatables Llanelli 01554 882000
Cosalt International Ltd Aberdeen 01224 826662
Glaslyn Marine Supplies Ltd Porthmadog 01766 513545
Hale Marine, Ron Portsmouth 023 9273 2985
Guernsey Yacht Club St Peter Port 01481 722838
IBS Boats South Woodham Ferrers 01245 323211/425551
KTS Seasafety Kilkeel 028 918 28405
Nationwide Marine Hire Warrington 01925 245788
Norwest Marine Ltd Liverpool 0151 207 2860
Ocean Safety Southampton 023 8072 0800
Polymarine Ltd Conwy 01492 583322
Premium Liferaft Services Burnham-on-Crouch 0800 243673
Ribeye Dartmouth 01803 832060
South Eastern Marine Services Ltd Basildon 01268 534427
Suffolk Marine Safety Ipswich 01473 833010
Whitstable Marine Whitstable 01227 262525

MARINAS

Aberystwyth Marina 01970 611422
Amble Marina 01665 712168
Arbroath Harbour 01241 872166
Ardfern Yacht Centre Ltd 01852 500247
Ardglass Marina 028 44842332
Arklow Marina +353 87 258 8078
Ballycastle Marina 028 2076 8525
Banff Harbour Marina 01261 832236
Bangor Marina 028 91 453297
Beaucette Marina 01481 245000
Bembridge Harbour Authority 01983 872828
Berthon Lymington Marina 01590 647405
Birdham Pool Marina 01243 512310
Blackwater Marina 01621 740264
Boston Gateway Marina 07480 525230
Bradwell Marina 01621 776235
Bray Marina 01628 623654
Brentford Dock Marina 020 8232 8941
Bridgemarsh Marine 01621 740414
Brighton Marina 01273 819919
Bristol Marina 0117 921 3198
Brixham Marina 01803 882929
Bucklers Hard Marina 01590 616200
Burnham Yacht Harbour Marina Ltd 01621 782150
Cahersiveen Marina +353 66 947 2777
Caley Marina 01463 236539
Campbeltown Marina 07798 524821
Cardiff Marina 02920 396078
Carlingford Marina +353 42 9373072
Carrickfergus Marina 028 9336 6666
Castlepark Marina +353 21 477 4959
Chatham Maritime Marina 01634 899200
Chelsea Harbour Marina 07770 542783
Chichester Marina 01243 512731
Clyde Marina Ltd 01294 607077
Cobbs Quay Marina 01202 674299
Coleraine Harbour Town Centre Marina 028 7034 2012
Coleraine Marina 028 703 44768

Conwy Marina 01492 593000
Cork Harbour Marina +353 87 3669009
Cowes Harbour Shepards Marina 01983 297821
Cowes Yacht Haven 01983 299975
Craobh Marina 01852 500222
Crinan Boatyard 01546 830232
Crosshaven Boatyard Marina +353 21 483 1161
Dart Marina Yacht Harbour 01803 837161
Darthaven Marina 01803 752242
Dartside Quay 01803 845445
Deacons Marina & Boatyard 02380 402253
Deganwy Marina 01492 576888
Dingle Marina +353 (0)87 925 4115
Douglas Marina 01624 686627
Dover Marina 01304 241663
Dun Laoghaire Marina +353 1 202 0040
Dunstaffnage Marina Ltd 01631 566555
East Cowes Marina 01983 293983
Emsworth Yacht Harbour 01243 377727
Essex Marina 01702 258531
Falmouth Haven Marina 01326 310991
Falmouth Marina 01326 316620
Fambridge Yacht Haven 01621 740370
Fambridge Yacht Station 01621 742911
Fenit Harbour & Marina +353 66 7136231
Fleetwood Haven Marina 01253 879062
Fox's Marina & Boatyard 01473 689111
Foyle Port Marina 02871 860555
Gallions Point Marina 0207 476 7054
Galway Harbour Marina +353 91 561874
Gillingham Marina 01634 280022
Glasson Waterside & Marina 01524 751491
Gosport Marina 023 9252 4811
Hafan Pwllheli 01758 701219
Hamble Point Marina 02380 452464
Harbour of Rye 01797 225225
Hartlepool Marina 01429 865744
Haslar Marina 023 9260 1201
Heybridge Basin 07712 079764
Holy Loch Marina 01369 701800
Holyhead Marina 01407 764242
Howth Marina +353 1839 2777
Hull Waterside & Marina 01482 609960
Humber Cruising Association 01472 268424
Hythe Marina Village 02380 207073
Inverness Marina 01463 220501
Ipswich Haven Marina 01473 236644
Island Harbour Marina 01983 539994
James Watt Dock Marina 01475 729838
Kemps Quay 023 8063 2323
Kilmore Quay Marina +353 53 91 29955
Kilrush Marina +353 65 9052072
Kinsale Yacht Club Marina +353 876 787377
Kip Marina 01475 521485
Kirkcudbright Marina 01557 331135
Kirkwall Marina 01856 871313
Lady Bee Marina 01273 593801
Lake Yard Marina 01202 674531
Largs Yacht Haven 01475 675333
Lawrence Cove Marina +353 27 75044
Limehouse Waterside & Marina 020 7308 9930
Littlehampton Marina 01903 713553
Liverpool Marina Bar & Grill 0151 707 6777
Lossiemouth Marina 01343 813066
Lowestoft Cruising Club 07810 522515
Lowestoft Haven Marina 01502 580300
Lymington Harbour Commission 01590 672014
Lymington Town Quay 01590 672014
Lymington Yacht Haven 01590 677071
Malahide Marina +353 1 845 4129
Mallaig Marina 01687 462406
Maryport Harbour and Marina 01900 814431
Mayflower International Marina 01752 556633
Melfort Pier & Harbour 01852 200333
Mercury Yacht Harbour and Holiday Park 023 8045 5994
Milford Marina 01646 696312
Mylor Yacht Harbour 01326 372121
Nairn Marina 01667 456008
Neptune Marina Ltd 01473 215204
New Ross Marina +353 86 3889652
Newhaven Marina 01273 513881
Neyland Yacht Haven 01646 601601
Northney Marina 02392 466321
Noss on Dart Marina 01803 839087
Oban Marina & Yacht Services Ltd 01631 565333
Ocean Village Marina 023 8022 9385
Padstow Harbour 01841 532239
Parkstone Yacht Club Haven 01202 738824
Peel Marina 01624 842338
Penarth Marina 02920 705021
Penton Hook 01932 568681
Peterhead Bay Marina 01779 477868
Plymouth Yacht Haven 01752 404231
Poole Quay Boat Haven 01202 649488
Port Bannatyne Marina 01700 503116
Port Edgar Marina 0131 331 3330
Port Ellen Marina 07464 151200
Port Hamble Marina 023 8045 2741
Port of Poole Marina 01202 649488
Port Pendennis Marina 01326 211211
Port Solent Marina 02392 210765
Port Werburgh 01634 252107
Portaferry Marina 07703 209780
Portavadie Marina 01700 811075
Portishead Marina 01275 841941
Portland Marina 01305 866190
Preston Marina 01772 733595
Queen Anne's Battery 01752 671142
Rhu Marina 01436 820238
Ridge Wharf Yacht Centre 01929 552650
Royal Clarence Marina 02392 523523
Royal Cork Yacht Club Marina +353 21 483 1023
Royal Harbour Marina, Ramsgate 01843 572100
Royal Harwich Yacht Club Marina 01473 780319
Royal Norfolk and Suffolk Yacht Club 01502 566726
Royal Northumberland Yacht Club 01670 353636
Royal Quays Marina 0191 272 8282
Ryde Leisure Harbour 01983 613879
Salterns Marina Ltd 01202 709971
Salve Engineering Marina +353 21 483 1145
Sandpoint Marina (Dumbarton) 01389 762396
Saxon Wharf 023 8033 9490
Seaport Marina 01463 725500
Seaton's Marina 028 703 832086
Shamrock Quay 023 8022 9461
Shotley Marina 01473 788982
South Dock Marina 020 7252 2244
South Ferriby Marina 01652 635620
Southsea Marina 02392 822719
Sovereign Harbour Marina 01323 470099

Sparkes Marina 023 92463572
St Helier Marina 01534 447708
St Katharine Docks Marina 0207 264 5312
St Peter Port Marinas 01481 720229
St Peter's Marina 0191 265 4472
Stornoway Marina 01851 702688
Stranraer Marina 01776 706565
Stromness Marina 01856 871313
Suffolk Yacht Harbour Ltd 01473 659240
Sunderland Marina 0191 514 4721
Sutton Harbour 01752 204702
Swansea Marina 01792 470310
Swanwick Marina 01489 884081
Tarbert Harbour 01880 820344
Titchmarsh Marina 01255 672185
Tobermory Hbr Auth 01688 302876
Tollesbury Marina 01621 869202
Torquay Marina 01803 200210
Town Quay Marina 02380 234397
Troon Yacht Haven 01292 315553
Universal Marina 01489 574272
Victoria Marina 01481 725987
Walton & Frinton Yacht Trust Limited 01255 675873
Waterford City Marina +353 87 238 4944
Weymouth Harbour 01305 838386
Weymouth Marina 01305 767576
Whitby Marina 01947 602354
Whitehaven Marina 01946 692435
Whitehills Marina 01261 861291
Wick Marina 01955 602030
Wicor Marine Yacht Haven 01329 237112
Windsor Marina 01753 853911
Wisbech Yacht Harbour 01945 588059
Woolverstone Marina and Lodge Park 01473 780206
Yarmouth Harbour 01983 760321

MARINE ENGINEERS

AAS Marine
Aberystwyth 01970 631090
Allerton Engineering
Lowestoft 01502 537870
APAS Engineering Ltd
Southampton 023 8063 2558
Ardmair Boat Centre
Ullapool 01854 612054
Arisaig Marine
Inverness-shire 01687 450224
Arun Craft
Littlehampton 01903 723667
ASL Auto Services
Boston 01205 761560
Atlantis Marine Power Ltd
Plymouth 01752 208810
Attrill & Sons, H
Bembridge 01983 872319
Auto & Marine Services
Botley 07836 507000
Auto Marine
Southsea 023 9282 5601
Baker, Keith
Brentford 07792 937790
BJ Marine Ltd Bangor 028 9127 1434
Bristol Boat Ltd
Bristol 01225 872032
Browne, Jimmy
Tralee +353 87 262 7158
Buccaneer Ltd
Macduff 01261 835199
Buzzard Marine Engineering
Yarmouth 01983 760707
C & B Marine Ltd
Chichester Marina 01243 511273
Caddy, Simon
Falmouth Marina
Falmouth 01326 372682
Caledonian Marine
Rhu Marina 01436 821184
Cardigan Outboards
Cardigan 01239 613966
Channel Islands Marine Ltd
Guernsey 01481 716880
Channel Islands Marine Ltd
Jersey 01534 767595
Cook's Diesel Service Ltd
Faversham 01795 538553
Cragie Engineering
Kirkwall 01856 874680
Crinan Boatyard Ltd
Crinan 01546 830232
Cutler Marine Engineering, John
Emsworth 01243 375014
Dale Sailing Co Ltd
Milford Haven 01646 603110
Davis Marine Services
Ramsgate 01843 586172
Denney & Son, EL
Redcar 01642 483507
DH Marine (Shetland) Ltd
Shetland 01595 690618
Dobson, Chris
Whitehaven 07986 086641
Emark Marine Ltd
Emsworth 01243 375383
Evans, Lyndon
Brentford 07795 218704
Evans Marine Engineering, Tony
Pwllheli 01758 703070
Ferrypoint Boat Co
Youghal +353 24 94232
Fettes & Rankine Engineering
Aberdeen 01224 573343
Floetree Ltd
Loch Lomond Marina
Balloch 01389 752069
Fowey Harbour Marine Engineers
Fowey 01726 832806
Fox Marine Services Ltd
Jersey 01534 721312
Freeport Marine Jersey 01534 888100
French Marine Motors Ltd
Colchester 01206 302133
French Marine Motors Ltd
Titchmarsh Marina 01255 850303
GH Douglas Marine Services
Fleetwood Harbour Village Marina,
Fleetwood 01253 877200
Golden Arrow Marine
Southampton 023 8071 0371
Goodchild Marine Services
Great Yarmouth 01493 782301
Goodwick Marine
Fishguard 01348 873955
Gosport Marina 023 9252 4811
Griffins Garage Dingle Marina,
Co Kerry +353 66 91 51178
Hale Marine, Ron
Portsmouth 023 9273 2985
Hamilton Brothers
Campbeltown 01586 553031
Hamnavoe Engineering
Stromness 01856 850576
Harbour Engineering
Itchenor 01243 513454
Hartlepool Marine Engineering
Hartlepool 01429 867883
Hayles, Harold
Yarmouth 01983 760373
Herm Seaway Marine Ltd
St Peter Port 01481 726829
HNP Engineers (Lerwick Ltd)
Lerwick 01595 692493
Hodges, T Coleraine 028 7035 6422
Home Marine Emsworth Yacht
Harbour, Emsworth 01243 374125
Hook Marine Ltd
Troon 01292 679500

Humphrey, Chris
Teignmouth 01626 772324

Instow Marine Services
Bideford 01271 861081

Jones (Boatbuilders), David
Chester 01244 390363

Keating Marine Engineering Ltd, Bill
Jersey 01534 733977

Kingston Marine Services
Cowes 01983 299385

Kippford Slipway Ltd
Dalbeattie 01556 620249

L&A Marine
Campbeltown 01586 554479

Lansdale Pannell Marine
Chichester 01243 550042

Lencraft Boats Ltd
Dungarvan +353 58 68220

Llyn Marine Services
Pwllheli 01758 612606

Lynx Engineering
St Helens, Isle of Wight 01983 873711

M&G Marine Services
Mayflower International Marina, Plymouth 01752 563345

MacDonald & Co Ltd, JN
Glasgow 0141 810 3400

Mackay Marine Services
Aberdeen 01224 575772

Mainbrayce Marine
Alderney 01481 722772

Malakoff and Moore
Lerwick 01595 695544

Marine Blast
Holy Loch 01369 705394

Mallaig Boat Building and Engineering
Mallaig 01687 462304

Marindus Engineering
Kilmore Quay +353 53 29794

Marine Engineering Looe
Brixham 01803 844777

Marine Engineering Looe
01503 263009

Marine Engineering Services
Port Dinorwic 01248 671215

Marine General Engineers Beaucette Marina, Guernsey 01481 245808

Marine Propulsion
Hayling Island 07836 737488

Marine & General Engineers
St. Sampsons Harbour, Guernsey 01481 245808

Marine Servicing
Eastbourne 07932 318414

Marine-Trak Engineering
Mylor Yacht Harbour 01326 376588

Marine Warehouse
Gosport 023 9258 0420

Marlec Marine
Ramsgate 01843 592176

Martin Outboards
Galgate 01524 751750

McKenzie, D
Campbeltown 07799 651637

McQueen, Michael
Kilrush +35 87 2574623

Meiher, Denis
Fenit +353 87 958 4744

MES Marine Greenock 01475 744655

MMS Ardrossan 01294 604831/ 07836 342332

Mobile Marine Engineering Liverpool Marina, Liverpool 01565 733553

Mobile Marine Maintenance
Poole 07931 776482

Mount's Bay Engineering
Newlyn 01736 363095

MP Marine Maryport 01900 810299

Murphy Marine Services
Cahersiveen +353 66 9476365

New World Yacht Care
Helensburgh 01436 820586

North Western Automarine Engineers
Largs 01475 687139

Noss Marine Services Dart Marina, Dartmouth 01803 833343

Owen Marine, Robert
Porthmadog 01766 513435

Pace, Andy
Newhaven 01273 516010

Penzance Dry Dock and Engineering Co Ltd Penzance 01736 363838

Pirie & Co, John S
Fraserburgh 01346 513314

Portavon Marine
Keynsham 01225 424301

Power Afloat, Elkins Boatyard
Christchurch 01202 489555

Powerplus Marine Cowes Yacht Haven, Cowes 01983 290421

Pro-Marine Queen Anne's Battery Marina, Plymouth 01752 267984

PT Marine Engineering
Hayling Island 023 9246 9332

R & M Marine
Portsmouth 023 9273 7555

R & S Engineering
Dingle Marina +353 66 915 1189

Reddish Marine
Salcombe 01548 844094

RHP Marine Cowes 01983 290421

RK Marine Ltd Hamble 01489 583585

RK Marine Ltd
Swanwick 01489 583572

Rossiter Yachts Ltd
Christchurch 01202 483250

Ryan & Roberts Marine Services
Askeaton +353 61 392198

Salve Marine Ltd
Crosshaven +353 21 4831145

Seamark-Nunn & Co
Felixstowe 01394 275327

Seapower
Ipswich 01473 780090

Seaward Engineering
Glasgow 0141 632 4910

Seaway Marine
Gosport 023 9260 2722

Shearwater Engineering Services Ltd
Dunoon 01369 706666

Silvers Marina Ltd
Helensburgh 01436 831222

Starey Marine
Salcombe 01548 843655

Swordfish Marine Engineering
Holy Loch 01369 701905

Tarbert Marine
Arbroath 01241 872879

Thorne Boat Services
Thorne 01405 814197

Tollesbury Marine Engineering
Tollesbury Marina 01621 869919

Tony's Marine Services
Coleraine 07866 690436

TOR (Gerald Hales)
Stornoway 01851 871025

Vasey Marine Engineering, Gordon
Fareham 07798 638625

Volspec Ltd Tollesbury 01621 869756

Wallis, Peter Torquay Marina, Torquay 01803 844777

Wartsila Havant 023 9240 0121

WB Marine Chichester 01243 512857

West Coast Marine
Troon 01292 318121

West Marine
Brighton 01273 626656

Weymouth Marina Mechanical Services Weymouth 01305 779379

Whittington, G Lady Bee Marine, Shoreham 01273 593801

Whitewater Marine Malahide +353 1 816 8473

Wigmore Wright Marine Services Penarth Marina 029 2070 9983

Wright, M Manaccan 01326 231502

Wyko Industrial Services Aberdeen 01224 246560

Ynys Marine Cardigan 01239 613179

Youngboats Faversham 01795 536176

1° West Marine Ltd Portsmouth 023 9283 8335

MASTS, SPARS & RIGGING

JWS Marine Services Portsmouth 02392 755155

A2 Rigging Falmouth 01326 312209

Allspars Plymouth 01752 266766

Amble Boat Co Ltd Morpeth 01665 710267

Arun Canvas & Rigging Littlehampton 01903 732561

B+ St Peter Port 01481 726071

Buchanan, Keith St Mary's 01720 422037

Bussell & Co, WL Weymouth 01305 785633

Carbospars Ltd Hamble 023 8045 6736

Cable & Rope Works Bexhill-on-Sea 01424 220112

Clarke Rigging, Niall Coleraine 07916 083858

Clyde Rigging Ardrossan 07773 244821

Coates Marine Ltd Whitby 01947 604486

Cunliffe, Alistair Fleetwood 07555 798310

Dauntless Boatyard Ltd Canvey Island 01268 793782

Davies Marine Services Ramsgate 01843 586172

Eurospars Ltd Plymouth 01752 550550

Exe Leisure Exeter 01392 879055

Fox's Marine Ipswich Ltd Ipswich 01473 689111

Freeland Yacht Spars Ltd Dorchester on Thames 01865 341277

Gordon, AD Portland 01305 821569

Hamble Custom Rigging Centre Hamble 023 8045 2000

Harris Rigging Totnes 01803 840160

Heyn Engineering Belfast 028 9035 0022

Holman Rigging Chichester 01243 514000

Irish Spars and Rigging Malahide +353 86 209 5996

JWS Marine Services Portsmouth 02392 755155

Kildale Marine Hull 01482 227464

Kilrush Marina Boatyard Kilrush +35 87 7990091

Lowestoft Yacht Services Lowestoft 01502 585535

Laverty, Billy Galway +353 86 3892614

Leitch, WB Tarbert, Loch Fyne 01880 820287

Lewis, Harry Kinsale +353 87 266 7127

Marine Resource Centre Oban 01631 720291

Martin Leaning Masts & Rigging Hayling 023 9237 1157

Mast & Rigging Crosshaven +353 21 483 3878

Mast & Rigging Services Largs 01475 670110

Mast & Rigging Services Inverkip 01475 522700

MP Marine Maryport 01900 810299

Ocean Rigging Lymington 01590 676292

Owen Sails Oban 01631 720485

Pro Rig S Ireland +353 87 298 3333

Ratsey, Stephen Milford Haven 01646 601561

Riglt Ardrossan 07593 220213

Rig Magic Ipswich 01473 655089

Rig Shop Southampton 023 8033 8341

Ronstan Gosport 023 9252 5377

Salcombe Boatstore Salcombe 01548 843708

Seldén Mast Ltd Gosport 01329 504000

Silvers Marina Ltd Helensburgh 01436 831222

Silverwood Yacht Services Ltd Portsmouth 023 9232 7067

Spencer Rigging Cowes 01983 292022

Storrar Marine Store Newcastle upon Tyne 0191 266 1037

Tedfords Rigging & Rafts Belfast 028 9032 6763

TJ Rigging Conwy 07780 972411

TS Rigging Maldon 01621 874861

Windjammer Marine Milford Marina 01646 699070

Yacht Rigging Services Plymouth 01752 226609

Z Spars UK Hadleigh 01473 822130

NAVIGATION EQUIPMENT – GENERAL

Belson Design Ltd, Nick Southampton 077 6835 1330

Brown Son & Ferguson Ltd Glasgow 0141 429 1234

Cooke & Son Ltd, B Hull 01482 223454

Diverse Yacht Services Hamble 023 8045 3399

Dolphin Maritime Software Ltd Lancaster 01524 841946

Dubois Phillips & McCallum Ltd Liverpool 0151 236 2776

Garmin Southampton 02380 524000

Geonav UK Ltd Poole 0870 240 4575

Imray Laurie Norie and Wilson Ltd St Ives, Cambs 01480 462114

Kelvin Hughes Southampton 023 8063 4911

Lilley & Gillie Ltd, John North Shields 0191 257 2217

Marine Chart Services Wellingborough 01933 441629

Navico UK Romsey 01794 510010

PC Maritime Plymouth 01752 254205

Price & Co, WF Bristol 0117 929 2229

Raymarine Ltd Portsmouth 023 9269 3611

Royal Institute of Navigation London 020 7591 3130

Sea Chest Nautical Bookshop
Plymouth 01752 222012

Seath Instruments (1992) Ltd
Lowestoft 01502 573811

Smith (Marine) Ltd, AM
London 020 8529 6988

South Bank Marine Charts Ltd
Grimsby 01472 361137

Southcoasting Navigators
Devon 01626 335626

Stanford Charts
Bristol 0117 929 9966
London 020 7836 1321
Manchester 0870 890 3730

Todd Chart Agency Ltd
County Down 028 9146 6640

UK Hydrographic Office
Taunton 01823 337900

Warsash Nautical Bookshop
Warsash 01489 572384

Yachting Instruments Ltd
Sturminster Newton 01258 817662

PAINT & OSMOSIS

Advanced Blast Cleaning Paint
Tavistock 01822 617192
07970 407911

Herm Seaway Marine Ltd
St Peter Port 01481 726829

Gillingham Marina 01634 280022

Hempel Paints
Southampton 02380 232000

International Coatings Ltd
Southampton 023 8022 6722

Marineware Ltd
Southampton 023 8033 0208

NLB Marine
Ardrossan 01563 521509

Pro-Boat Ltd
Burnham on Crouch 01621 785455

Rustbuster Ltd
Peterborough 01775 761222

SP Systems
Isle of Wight 01983 828000

PROPELLERS & STERGEAR/REPAIRS

CJR Propulsion Ltd
Southampton 023 8063 9366

Darglow Engineering Ltd
Wareham 01929 556512

Propeller Revolutions
Poole 01202 671226

Sillette – Sonic Ltd
Sutton 020 8337 7543

Vetus Den Ouden Ltd
Southampton 02380 454507

RADIO COURSES / SCHOOLS

Bisham Abbey Sailing & Navigation School Bisham 01628 474960

East Coast Offshore Yachting – Les Rant Perry 01480 861381

Hamble School of Yachting
Hamble 023 8045 6687

Pembrokeshire Cruising
Neyland 01646 602500

Plymouth Sailing School
Plymouth 01752 493377

Start Point Sailing
Kingsbridge 01548 810917

REEFING SYSTEMS

Atlantic Spars Ltd
Brixham 01803 843322

Calibra Marine International Ltd
Southampton 08702 400358

Eurospars Ltd
Plymouth 01752 550550

Holman Rigging
Chichester 01243 514000

Sea Teach Ltd
Emsworth 01243 375774

Southern Spar Services
Northam 023 8033 1714

Wragg, Chris
Lymington 01590 677052

Z Spars UK Hadleigh 01473 822130

REPAIR MATERIALS & ACCESSORIES

Akeron Ltd
Southend on Sea 01702 297101

Howells & Son, KJ Poole 01202 665724

JB Timber Ltd
North Ferriby 01482 631765

Robbins Timber Bristol 0117 9633136

Sika Ltd
Welwyn Garden City 01707 394444

Solent Composite Systems
East Cowes 01983 292602

Technix Rubber & Plastics Ltd
Southampton 01489 789944

Tiflex Liskeard 01579 320808

Timage & Co Ltd
Braintree 01376 343087

Trade Grade Products Ltd
Poole 01202 820177

Wessex Resins & Adhesives Ltd
Romsey 01794 521111

ROPE & WIRE

Cable & Rope Works
Bexhill-on-Sea 01424 220112

Euro Rope Ltd
Scunthorpe 01724 280480

Marlow Ropes
Hailsham 01323 444444

Mr Splice Leicester 0800 1697178

Spinlock Ltd Cowes 01983 295555

TJ Rigging Conwy 07780 972411

SAFETY EQUIPMENT

AB Marine Ltd
St Peter Port 01481 722378

Adec Marine Ltd
Croydon 020 8686 9717

Anchorwatch UK
Edinburgh 0131 447 5057

Avon Inflatables
Llanelli 01554 882000

Cosalt International Ltd
Aberdeen 01224 588327

Crewsaver Gosport 01329 820000

Glaslyn Marine Supplies Ltd
Porthmadog 01766 513545

Exposure Lights
Pulborough 01798 839300

Guardian Fire Protection
Manchester 0800 358 7522

Hale Marine, Ron
Portsmouth 023 9273 2985

Herm Seaway Marine Ltd
St Peter Port 01481 722838

KTS Seasafety Kilkeel 028 41762655

McMurdo Pains Wessex
Portsmouth 023 9262 3900

Met Office Exeter 0870 900 0100

Nationwide Marine Hire
Warrington 01925 245788

Norwest Marine Ltd
Liverpool 0151 207 2860

Ocean Safety
Southampton 023 8072 0800

Polymarine Ltd
Conwy 01492 583322

Premium Liferaft Services
Burnham-on-Crouch 0800 243673

Ribeye Dartmouth 01803 832060

South Eastern Marine Services Ltd
Basildon 01268 534427

Suffolk Sailing
Ipswich 01473 604678

Whitstable Marine
Whitstable 01227 262525

Winters Marine Ltd
Salcombe 01548 843580

SAILMAKERS & REPAIRS

Alsop Sailmakers, John
Salcombe 01548 843702

AM Trimming
Windsor 01932 821090

Arun Canvas & Rigging
Littlehampton 01903 732561

Barrett, Katy
St Peter Port 07781 404299

Batt Sails Bosham 01243 575505

Bissett and Ross
Aberdeen 01224 580659

Boatshed, The
Felinheli, Bangor 01248 679939

Breaksea Sails Barry 01446 730785

Bristol Sails Bristol 0117 922 5080

Buchanan, Keith
St Mary's 01720 422037

C&J Marine Ltd
Chichester 01243 785485

Calibra Sails
Dartmouth 01803 833094

Clarke Rigging, Niall
Coleraine 07916 083858

Coastal Covers
Portsmouth 023 9252 0200

Covercare Fareham 01329 311878

Covers + Stuff
Douglas, Isle of Man 07624 400037

Crawford, Margaret
Kirkwall 01856 875692

Crusader Sails Poole 01202 670580

Crystal Covers
Portsmouth 023 9238 0143

Cullen Sailmakers
Galway +353 91 771991

Dolphin Sails Harwich 01255 243366

Doyle Sails
Southampton 023 8033 2622

Downer International Sails & Chandlery
Dun Laoghaire +353 1 280 0231

Duthie Marine Safety, Arthur
Glasgow 0141 429 4553

Dynamic Sails
Emsworth 01243 374495

Flew Sailmakers
Portchester 01329 822676

Fylde Coast Sailmaking Co
Fleetwood 01253 873476

Freeman Sails Padstow 07771 610053

Garland Sails Bristol 01275 393473

Goacher Sails
Cumbria 01539 488686

Gowen Ocean Sailmakers
West Mersea 01206 384412

Green Sailmakers, Paul
Plymouth 01752 660317

Henderson Sails & Covers
Southsea 023 9229 4700

Hood Sailmakers UK
Wareham 0844 209 4789

Hooper, A Plymouth 01752 830411

Hyde Sails
Southampton 0845 543 8945

Jackson Yacht Services
Jersey 01534 743819

Jeckells and Son Ltd (Wroxham)
Wroxham 01603 782223

Jessail Ardrossan 01294 467311

JKA Sailmakers
Pwllheli 01758 613266

Kemp Sails Ltd
Wareham 01929 554308/554378

Kildale Marine Hull 01482 227464

Lawrence Sailmakers, J
Brightlingsea 01206 302863

Leitch, WB
Tarbert, Loch Fyne 01880 820287

Leith UK
Berwick on Tweed 01289 307264

Le Monnier, Yannick
Galway +353 87 628 9854

Lodey Sails
Newlyn 01736 719359

Lossie Sails
Lossiemouth 07989 956698

Lucas Sails Portchester 023 9237 3699

Malakoff and Moore
Lerwick 01595 695544

McCready and Co Ltd, J
Belfast 028 90232842

McKillop Sails, John
Kingsbridge 01548 852343

McNamara Sails, Michael
Great Yarmouth 01692 584186

McWilliam Sailmaker (Crosshaven)
Crosshaven +353 21 4831505

Sail Shape
Fowey 01726 833731

Montrose Rope and Sails
Montrose 01674 672657

Mountfield Sails
Hayling Island 023 9246 3720

Mouse Sails Holyhead 01407 763636

Nicholson Hughes Sails
Rosneath 01436 831356

North Sea Sails
Tollesbury 01621 869367

North West Sails
Keighley 01535 652949

Northrop Sails
Ramsgate 01843 851665

O'Mahony Sailmakers
Kinsale +353 86 326 0018

O'Sullivans Marine Ltd
Tralee +353 66 7129635

Owen Sails Benderloch 01631 720485

Parker & Kay Sailmakers –
East Ipswich 01473 659878

Parker & Kay Sailmakers –
South Hamble 023 8045 8213

Penrose Sailmakers
Falmouth 01326 312705

Pinnell & Bax
Northampton 01604 592808

Pollard Marine
Port St Mary 01624 835831

Quantum Sails
Ipswich Haven Marina 01473 659878

Quantum-Parker & Kay Sailmakers
Hamble 023 8045 8213

Quay Sails (Poole) Ltd
Poole 01202 681128

Ratsey & Lapthorn
Isle of Wight 01983 294051

Ratsey Sailmakers, Stephen
Milford Haven 01646 601561

Relling One Design
Portland 01305 826555

SO31 Bags
Southampton 023 8045 5106

Rig Shop, The
Southampton 023 8033 8341

Rockall Sails Chichester 01243 573185

Sail Locker
Woolverstone Marina 01473 780206

Sail Style
Hayling Island 023 9246 3720

Sails & Canvas Exeter 01392 877527

Sail Register Ulceby 01469 589444

Saltern Sail Co
West Cowes 01983 280014

Saltern Sail Company
Yarmouth 01983 760120

Sanders Sails
Lymington 01590 673981

Saturn Sails Largs 01475 689933

Scott & Co, Graham
St Peter Port 01481 259380

Shore Sailmakers
Swanwick 01489 589450

SKB Sails Falmouth 01326 372107

Sketrick Sailmakers Ltd
Killinchy 028 9754 1400

Solo Sails Penzance 01736 366004

Storrar Marine Store
Newcastle upon Tyne 0191 266 1037

Suffolk Sails
Woodbridge 01394 386323

Sunset Sails Sligo +353 71 62792

Torquay Marina Sails and Canvas
Exeter 01392 877527

Trident UK Gateshead 0191 490 1736

Sailcare (UK) Ltd Cowes 01983 248589

W Sails Leigh-on-Sea 01702 714550

Warren Hall Beaucette 07781 444280

Watson Sails Dublin +353 1 846 2206

WB Leitch and Son
Tarbert 01880 820287

Westaway Sails
Plymouth Yacht Haven 01752 892560

Westsails.ie Kilrush +35 876289854

Wilkinson Sails
Burnham-on-Crouch 01621 786770

Wilkinson Sails
Teynham 01795 521503

Yacht Shop, The
Fleetwood 01253 879238

SOLAR POWER

Ampair Ringwood 01425 480780

Barden UK Ltd Fareham 01489 570770

Marlec Engineering Co Ltd
Corby 01536 201588

SPRAYHOODS & DODGERS

A & B Textiles
Gillingham 01634 579686

Allison–Gray Dundee 01382 505888

Arton, Charles
Milford-on-Sea 01590 644682

Arun Canvas and Rigging Ltd
Littlehampton 01903 732561

Boatshed, The
Felinheli, Bangor 01248 679939

Buchanan, Keith
St Mary's 01720 422037

C & J Marine Textiles
Chichester 01243 785485

Covercare Fareham 01329 311878

Covercraft
Southampton 023 8033 8286

Crystal Covers
Portsmouth 023 9238 0143

Forth Marine Textiles
Holy Loch 01383 622444

Jeckells and Son Ltd
Wroxham 01603 782223

Jessail Ardrossan 01294 467311

Lomond Boat Covers
Alexandria 01389 602734

Lucas Sails
Portchester 023 9237 3699

Poole Canvas Co Ltd
Poole 01202 677477

Sail Register Ulceby 01469 589444

Saundersfoot Auto Marine
Saundersfoot 01834 812115

Trident UK
Gateshead 0191 490 1736

SURVEYORS AND NAVAL ARCHITECTS

Amble Boat Company Ltd
Amble 01665 710267

Ark Surveys East Anglia/South Coast
01621 857065/01794 521957

Atkin & Associates
Lymington 01590 688633

Barbican Yacht Agency Ltd
Plymouth 01752 228855

Battick, Lee St Helier 01534 611143

Booth Marine Surveys, Graham
Birchington-on-Sea 01843 843793

Byrde & Associates
Kimmeridge 01929 480064

Bureau Maritime Ltd
Maldon 01621 859181

Byrde & Associates
Kimmeridge 01929 480064

Cannell & Associates, David M
Wivenhoe 01206 823337

Cardiff Commercial Boat Operators Ltd
Cardiff 029 2037 7872

Clarke Designs LLP, Owen
Dartmouth 01803 770495

Connor, Richard
Coleraine 07712 115751

Cox, David Penryn 01326 340808

Davies, Peter N
Wivenhoe 01206 823289

Down Marine Co Ltd
Belfast 028 90480247

Evans, Martin
Kirby le Soken 07887 724055

Goodall, JL Whitby 01947 604791

Green, James Plymouth 01752 660516

Greening Naval Architect Ltd, David
Salcombe 01548 842000

Hansing & Associates
North Wales/Midlands 01248 671291

JP Services – Marine Safety & Training
Chichester 01243 537552

MacGregor, WA
Felixstowe 01394 676034

Mahoney & Co, KPO
Co Cork +353 21 477 6150

Marinte Surveys UK
Emsworth 07798 554535

Marintec Lymington 01590 683414

Norwood Marine
Margate 01843 835711

Quay Consultants Ltd
West Wittering 01243 673056

Scott Marine Surveyors & Consultants
Conwy 01492 573001

S Roberts Marine Ltd
Liverpool 0151 707 8300

Staton-Bevan, Tony
Lymington 01590 645755/
07850 315744

Thomas, Stephen
Southampton 023 8048 6273

Towler, Perrin
Lymington 01590 718087

Victoria Yacht Surveys 0800 083 2113

Ward & McKenzie
Woodbridge 01394 383222

Ward & McKenzie (North East)
Pocklington 01759 304322

YDSA Yacht Designers & Surveyors Association Bordon 0845 0900162

TAPE TECHNOLOGY

CC Marine Services (Rubbaweld) Ltd
London 020 7402 4009

Trade Grade Products Ltd
Poole 01202 820177

UK Epoxy Resins
Burscough 01704 892364

3M United Kingdom plc
Bracknell 0870 5360036

TRANSPORT/YACHT DELIVERIES

Boat Shifters
07733 344018/01326 210548

Convoi Exceptionnel Ltd
Hamble 023 8045 3045

Debbage Yachting
Ipswich 01473 601169

East Coast Offshore Yachting
01480 861381

Hainsworth's UK and Continental
Bingley 01274 565925

Houghton Boat Transport
Tewkesbury 07831 486710

MCL Transboat 08455 201900

MJS Boat Transport
Alexandria, Scotland 01389 755047

Moonfleet Sailing
Poole 01202 682269

Performance Yachting
Plymouth 01752 565023

Peters & May Ltd
Southampton 023 8048 0480

Reeder School of Seamanship, Mike
Lymington 01590 674560

Seafix Boat Transfer 01766 514507

Sealand Boat Deliveries Ltd
Liverpool 01254 705225

Shearwater Sailing
Southampton 01962 775213

Southcoasting Navigators
Devon 01626 335626

West Country Boat Transport
01566 785651

Wolff, David 07659 550131

TUITION/SAILING SCHOOLS

Association of Scottish Yacht Charterers Argyll 07787 363562
01852 200258

Bisham Abbey Sailing & Navigation School Bisham 01628 474960

Blue Baker Yachts
Ipswich 01473 780008

British Offshore Sailing School
Hamble 023 8045 7733

Coastal Sea School
Weymouth 0870 321 3271

Dart Harbour Sea School
Dartmouth 01803 839339

Dartmouth Sailing
Dartmouth 01803 833399

Drake Sailing School
Plymouth 01635 253009

East Anglian Sea School
Ipswich 01473 659992

East Coast Offshore Yachting – Les Rant Perry 01480 861381

Gibraltar Sailing Centre +350 78554

Glenans Irish Sailing School
Baltimore +353 28 20154

Hamble School of Yachting
Hamble 023 8045 6687

Haslar Sea School
Gosport 023 9260 2708

Hobo Yachting
Southampton 023 8033 4574

Hoylake Sailing School
Wirral 0151 632 4664

Ibiza Sailing School 07092 235 853

International Yachtmaster Academy
Southampton 0800 515439

JP Services – Marine Safety & Training
Chichester 01243 537552

Lymington Cruising School
Lymington 01590 677478

Marine Leisure Association (MLA)
Southampton 023 8029 3822

Menorca Cruising School
01995 679240

Moonfleet Sailing Poole 01202 682269

National Marine Correspondence School Macclesfield 01625 262365

Pembrokeshire Cruising
Neyland 01646 602500

Performance Yachting & Chandlery
Plymouth 01752 565023

Plain Sailing
Dartmouth 01803 853843

Plymouth Sailing School
Plymouth 01752 493377

Port Edgar Marina & Sailing School
Port Edgar 0131 331 3330

Portsmouth Outdoor Centre
Portsmouth 023 9266 3873

Portugal Sail & Power 01473 833001

Safe Water Training Sea School Ltd
Wirral 0151 630 0466

Sail East Felixstowe 01255 502887

Sail East
River Crouch 07860 271954

Sail East
River Medway 07932 157027

Sail East Harwich 01473 689344

Sally Water Training
East Cowes 01983 299033

Sea-N-Shore
Salcombe 01548 842276

Seafever
East Grinstead 01342 316293

Solaris Mediterranean Sea School
01925 642909

Solent School of Yachting
Southampton 023 8045 7733

Southcoasting Navigators
Devon 01626 335626

Southern Sailing
Southampton 01489 575511

Start Point Sailing
Dartmouth 01548 810917

Sunsail
Port Solent/Largs 0870 770 6314

Team Sailing Gosport 023 9252 4370

Workman Marine School
Portishead 01275 845844

Wride School of Sailing, Bob
North Ferriby 01482 635623

WATERSIDE ACCOMMODATION & RESTAURANTS

51st State Bar & Grill
Dunoon 01369 703595

Abbey, The Penzance 01736 366906

Al Porto
Hull Marina 01482 238889

Arun View Inn, The
Littlehampton 01903 722335

Baywatch on the Beach
Bembridge 01983 873259

Beaucette Marina Restaurant Guernsey 01481 247066
Bella Napoli Brighton Marina 01273 818577
Bembridge Coast Hotel Bembridge 01983 873931
Budock Vean Hotel Porth Navas Creek 01326 252100
Café Mozart Cowes 01983 293681
Caffé Uno Port Solent 023 9237 5227
Chandlers Bar & Bistro Queen Anne's Battery Marina Plymouth 01752 257772
Cruzzo Malahide Marina Co Dublin +353 1 845 0599
Cullins Yard Bistro Dover 01304 211666
Custom House, The Poole 01202 676767
Dart Marina River Lounge Dartmouth 01803 832580
Doghouse Swanwick Marina, Hamble 01489 571602
Dolphin Restaurant Gorey 01534 853370
Doune Knoydart 01687 462667
El Puertos Penarth Marina 029 2070 5551
Falmouth Marina Marine Bar and Restaurant Falmouth 01326 313481
Ferry Boat Inn West Wick Marina Nr Chelmsford 01621 740208
Ferry Inn, The (restaurant) Pembroke Dock 01646 682947
Fisherman's Wharf Sandwich 01304 613636
Folly Inn Cowes 01983 297171
Gaffs Restaurant Fenit, County Kerry +353 66 71 36666
Godleys Hotel Fenit County Kerry +353 66 71 36108
Harbour Lights Restaurant Walton on the Naze 01255 851887
Haven Bar and Bistro, The Lymington Yacht Haven 01590 679971
Haven Hotel Poole 08453 371550
HMS Ganges Restaurant Mylor Yacht Harbour 01326 374320
Holy Loch Inn Holy Loch 01369 706903
Hunters Bar & Grill Holy Loch 01369 707772
Jolly Sailor, The Bursledon 023 8040 5557
Kames Hotel Argyll 01700 811489
Ketch Rigger, The Hamble Point Marina Hamble 023 8045 5601
Kings Arms Stoborough 01929 552705
Kota Restaurant Porthleven 01326 562407
La Cala Lady Bee Marina Shoreham 01273 597422
La Cantina Restaurant Dunoon 01369 703595
Lake Yard Ltd Poole 01202 676953
Le Nautique St Peter Port 01481 721714
Lighter Inn, The Topsham 01392 875439
Mariners Bistro Sparkes Marina, Hayling Island 023 9246 9459
Mary Mouse II Haslar Marina, Gosport 023 9252 5200
Martha's Vineyard Milford Haven 01646 697083
Master Builder's House Hotel Buckler's Hard 01590 616253
Millstream Hotel Bosham 01243 573234
Montagu Arms Hotel Beaulieu 01590 612324
Oyster Quay Mercury Yacht Harbour, Hamble 023 8045 7220
Paris Hotel Coverack 01326 280258
Pebble Beach, The Gosport 023 9251 0789
Philip Leisure Group Dartmouth 01803 833351
Priory Bay Hotel Seaview, Isle of Wight 01983 613146
QC's Cahersiveen +353 66 9472244
Quay Bar Galway +353 91 568347
Quayside Hotel Brixham 01803 855751
Queen's Hotel Kirkwall 01856 872200
Sails Dartmouth 01803 839281
Shell Bay Seafood Restaurant Poole Harbour 01929 450363
Simply Italian Sovereign Harbour, Eastbourne 01323 470911
Spit Sand Fort The Solent 01329 242077
Steamboat Inn Lossiemouth 01343 812066
Tayvallich Inn, The Argyll 01546 870282
Villa Adriana Newhaven Marina Newhaven 01273 513976
Warehouse Brasserie, The Poole 01202 677238
36 on the Quay Emsworth 01243 375592

WEATHER INFORMATION

Met Office Exeter 0870 900 0100

WOOD FITTINGS

Howells & Son, KJ Poole 01202 665724
Onward Trading Co Ltd Southampton 01489 885250
Robbins Timber Bristol 0117 963 3136
Sheraton Marine Cabinet Witney 01993 868275

YACHT BROKERS

ABC Powermarine Beaumaris 01248 811413
ABYA Association of Brokers & Yacht Agents Bordon 0845 0900162
Adur Boat Sales Southwick 01273 596680
Ancasta International Boat Sales Southampton 023 8045 0000
Anglia Yacht Brokerage Bury St Edmunds 01359 271747
Ardmair Boat Centre Ullapool 01854 612054
Barbican Yacht Agency, The Plymouth 01752 228855
Bates Wharf Marine Sales Ltd 01932 571141
BJ Marine Bangor 028 9127 1434
Boatworks + Ltd St Peter Port 01481 726071
Caley Marina Inverness 01463 236539
Calibra Marine International Ltd Southampton 08702 400358
Camper & Nicholsons International London 020 7009 1950
Clarke & Carter Interyacht Ltd Ipswich/Burnham on Crouch 01473 659681/01621 785600
Coastal Leisure Ltd Southampton 023 8033 2222
Dale Sailing Brokerage Neyland 01646 603105
Deacons Southampton 023 8040 2253
Exe Leisure Topsham 001392 879055
Ferrypoint Boat Co Youghal +353 24 94232
Gweek Quay Boatyard Helston 01326 221657
International Barge & Yacht Brokers Southampton 023 8045 5205
Iron Wharf Boatyard Faversham 01795 536296
Jackson Yacht Services Jersey 01534 743819

Kings Yacht Agency
Beaulieu 01590 616316

Kippford Slipway Ltd
Dalbeattie 01556 620249

Lencraft Boats Ltd
Dungarvan +353 58 68220

Liberty Yachts Ltd
Plymouth 01752 227911

Lucas Yachting, Mike
Torquay 01803 212840

Network Yacht Brokers
Dartmouth 01803 834864

Network Yacht Brokers
Plymouth 01752 605377

New Horizon Yacht Agency
Guernsey 01481 726335

Oyster Brokerage Ltd
Southampton 02380 831011

Pearn and Co, Norman
(Looe Boatyard) Looe 01503 262244

Performance Boat Company
Maidenhead 07768 464717

Peters Chandlery
Chichester 01243 511033

Portavon Marina
Keynsham 0117 986 1626

Retreat BY Topsham 01392 874720

SD Marine Ltd
Southampton 023 8045 7278

Sea & Shore Ship Chandler
Dundee 01382 202666

South Pier Shipyard
St Helier 01534 711000

South West Yacht Brokers Group
Plymouth 01752 401421

Sunbird Marine Services
Fareham 01329 842613

Swordfish Marine Brokerage
Holy Loch 01369 701905

Trafalgar Yacht Services
Fareham 01329 823577

Transworld Yachts
Hamble 023 8045 7704

Walker Boat Sales
Deganwy 01492 555706

Walton Marine Sales
Brighton 01273 670707
Portishead 01275 840132
Wroxham 01603 781178

Watson Marine, Charles
Hamble 023 8045 6505

Western Marine
Dublin +353 1280 0321

Woodrolfe Brokerage
Maldon 01621 868494

Youngboats Faversham 01795 536176

YACHT CHARTERS & HOLIDAYS

Ardmair Boat Centre
Ullapool 01854 612054

Association of Scottish Yacht Charterers Argyll 01880 820012

Camper & Nicholsons International
London 020 7009 1950

Coastal Leisure Ltd
Southampton 023 8033 2222

Crusader Yachting
Turkey 01732 867321

Dartmouth Sailing
Dartmouth 01803 833399

Dartmouth Yacht Charters
Kingswear 01803 883501

Doune Marine Mallaig 01687 462667

Four Seasons Yacht Charter
Gosport 023 9251 1789

Golden Black Sailing
Cornwall 01209 715757

Hamble Point Yacht Charters
Hamble 023 8045 7110

Haslar Marina & Victory Yacht Charters
Gosport 023 9252 0099

Indulgence Yacht Charters
Padstow 01841 719090

Liberty Yachts West Country, Greece, Mallorca & Italy 01752 227911

Nautilus Yachting
Mediterranean & Caribbean
01732 867445

On Deck Sailing
Southampton 023 8063 9997

Patriot Charters & Sail School
Milford Haven 01437 741202

West Country Yachts
Plymouth 01752 606999
Falmouth 01326 212320

Puffin Yachts Port Solent01483 420728

Sailing Holidays Ltd
Mediterranean 020 8459 8787

Sailing Holidays in Ireland
Kinsale +353 21 477 2927

Setsail Holidays
Greece, Turkey, Croatia, Majorca 01787 310445

Shannon Sailing Ltd
Tipperary +353 67 24499

Sleat Marine Services
Isle of Skye 01471 844216

Smart Yachts
Mediterranean 01425 614804

South West Marine Training
Dartmouth 01803 853843

Sovereign Sailingl
Kinsale +353 87 6172555

Sunsail
Worldwide 0870 770 0102

Templecraft Yacht Charters
Lewes 01273 812333

TJ Sailing
Gosport 07803 499691

Top Yacht Sailing Ltd
Havant 02392 347655

Victory Yacht Charters
Gosport 023 9252 0099

West Wales Yacht Charter
Pwllheli 07748 634869

39 North (Mediterranean)
Kingskerwell 07071 393939

YACHT CLUBS

Aberaeron YC
Aberdovey 01545 570077

Aberdeen and Stonehaven SC
Nr Inverurie 01569 764006

Aberdour BC 01383 860029

Abersoch Power BC
Abersoch 01758 712027

Aberystwyth BC
Aberystwyth 01970 624575

Aldeburgh YC 01728 452562

Alderney SC 01481 822959

Alexandra YC
Southend-on-Sea 01702 340363

Arklow SC +353 402 33100

Arun YC
Littlehampton 01903 716016

Axe YC Axemouth 01297 20043

Ayr Yacht and CC 01292 476034

Ballyholme YC
Bangor 028 91271467

Baltimore SC +353 28 20426

Banff SC 01464 820308

Bantry Bay SC +353 27 51724

Barry YC 01446 735511

Beaulieu River SC
Brockenhurst 01590 616273

Bembridge SC
Isle of Wight 01983 872237

Benfleet YC
Canvey Island 01268 792278

Blackpool and Fleetwood YC 01253 884205

Blackwater SC Maldon 01621 853923

Blundellsands SC 0151 929 2101

Bosham SC Chichester 01243 572341

Brading Haven YC
Isle of Wight 01983 872289

Bradwell CC 01621 892970

Bradwell Quay YC
Wickford 01268 890173

Brancaster Staithe SC 01485 210249

Brandy Hole YC
Hullbridge 01702 230320

Brightlingsea SC 01206 303275

Brighton Marina YC
Peacehaven 01273 818711

Bristol Avon SC 01225 873472

Bristol Channel YC
Swansea 01792 366000

Bristol Corinthian YC
Axbridge 01934 732033

Brixham YC 01803 853332

Burnham Overy Staithe SC 01328 730961

Burnham-on-Crouch SC
01621 782812

Burnham-on-Sea SC
Bridgwater 01278 792911

Burry Port YC 01554 833635

Cabot CC 01275 855207

Caernarfon SC (Menai Strait)
Caernarfon 01286 672861

Campbeltown SC 01586 552488

Island YC Canvey Island 01702 510360

Cardiff YC 029 2046 3697

Cardiff Bay YC 029 20226575

Carlingford Lough YC
Rostrevor 028 4173 8604

Carrickfergus SC
Whitehead 028 93 351402

Castle Cove SC
Weymouth 01305 783708

Castlegate Marine Club
Stockton on Tees 01642 583299

Chanonry SC Fortrose 01463 221415

Chichester Cruiser and Racing Club
01483 770391

Chichester YC 01243 512918

Christchurch SC 01202 483150

Clyde CC Glasgow 0141 221 2774

Co Antrim YC
Carrickfergus 028 9337 2322

Cobnor Activities Centre Trust
01243 572791

Coleraine YC 028 703 44503

Colne YC Brightlingsea 01206 302594

Conwy YC Deganwy 01492 583690

Coquet YC 01665 710367

Corrib Rowing & YC
Galway City +353 91 564560

Cowes Combined Clubs
01983 295744

Cowes Corinthian YC
Isle of Wight 01983 296333

Cowes Yachting 01983 280770

Cramond BC 0131 336 1356

Creeksea SC
Burnham-on-Crouch 01245 320578

Crookhaven SC +353 87 2379997

Crouch YC
Burnham-on-Crouch 01621 782252

Dale YC 01646 636362

Dartmouth YC 01803 832305

Deben YC Woodbridge 01394 384440

Dell Quay SC Chichester 01243 514639

Dingle SC +353 66 51984

Douglas Bay YC 01624 673965

Dovey YC
Aberdovey 01213 600008

Dun Laoghaire MYC +353 1 288 938

Dunbar SC
Cockburnspath 01368 86287

East Antrim BC 028 28 277204

East Belfast YC 028 9065 6283

East Cowes SC 01983 531687

East Dorset SC Poole 01202 706111

East Lothian YC 01620 892698

Eastney Cruising Association
Portsmouth 023 92734103

Eling SC 023 80863987

Emsworth SC 01243 372850

Emsworth Slipper SC 01243 372523

Essex YC Southend 01702 478404

Exe SC (River Exe)
Exmouth 01395 264607

Eyott SC Mayland 01245 320703

Fairlie YC 01294 213940

Falmouth Town SC 01326 373915

Falmouth Watersports Association
Falmouth 01326 211223

Fareham Sailing & Motor BC
Fareham 01329 280738

Felixstowe Ferry SC 01394 283785

Findhorn YC Findhorn 01309 690247

Fishguard Bay YC
Lower Fishguard 01348 872866

Flushing SC Falmouth 01326 374043

Folkestone Yacht and Motor BC
Folkestone 01303 251574

Forth Corinthian YC
Haddington 0131 552 5939

Forth YCs Association
Edinburgh 0131 552 3006

Fowey Gallants SC 01726 832335

Foynes YC Foynes +353 69 91201

Galway Bay SC +353 91 794527

Glasson SC Lancaster 01524 751089

Glenans Irish Sailing School
+353 1 6611481

Glenans Irish SC (Westport)
+353 98 26046

Gosport CC Gosport 02392 586838

Gravesend SC 07538 326623

Greenwich YC London 020 8858 7339

Grimsby and Cleethorpes YC
Grimsby 01472 356678

Guernsey YC
St Peter Port 01481 725342

Hamble River SC
Southampton 023 80452070

Hampton Pier YC
Herne Bay 01227 364749

Hardway SC Gosport 023 9258 1875

Hartlepool YC 01429 233423

Harwich Town SC 01255 503200

Hastings and St Leonards YC
Hastings 01424 420656

Haven Ports YC
Woodbridge 01473 659658

Hayling Ferry SC; Locks SC
Hayling Island 07870 367571

Hayling Island SC 023 92463768

Helensburgh SC Rhu 01436 672778

Helensburgh 01436 821234

Helford River SC
Helston 01326 231006

Herne Bay SC 01227 375650

Highcliffe SC
Christchurch 01425 274874

Holyhead SC 01407 762526

Holywood YC 028 90423355

Hoo Ness YC Sidcup 01634 250052

Hornet SC Gosport 023 9258 0403

Howth YC +353 1 832 2141

Hoylake SC Wirral 0151 632 2616

Hullbridge YC 01702 231797

Humber Yawl Club 01482 667224

Hundred of Hoo SC 01634 250102

Hurlingham YC
London 020 8788 5547

Hurst Castle SC 01590 645589

Hythe SC Southampton	02380 846563
Hythe & Saltwood SC	01303 265178
Ilfracombe YC	01271 863969
Iniscealtra SC Limerick	+353 61 338347
Invergordon BC	01349 893772
Irish CC	+353 214870031
Island CC Salcombe	01548 844631
Island SC Isle of Wight	01983 296621
Island YC Canvey Island	01268 510360
Isle of Bute SC Rothesay	01700 502819
Isle of Man YC Port St Mary	01624 832088
Itchenor SC Chichester	01243 512400
Keyhaven YC	01590 642165
Killyleagh YC	028 4482 8250
Kircubbin SC	028 4273 8422
Kirkcudbright SC	01557 331727
Langstone SC Havant	023 9248 4577
Largs SC Largs	01475 670000
Larne Rowing & SC	028 2827 4573
Lawrenny YC	01646 651212
Leigh-on-Sea SC	01702 476788
Lerwick BC	01595 696954
Lilliput SC Poole	01202 740319
Littlehampton Yacht Club Littlehampton	01903 713990
Loch Ryan SC Stranraer	01776 706322
Lochaber YC Fort William	01397 772361
Locks SC Portsmouth	07980 856267
Looe SC	01503 262559
Lossiemouth CC Fochabers	01348 812121
Lough Swilly YC Fahn	+353 74 22377
Lowestoft CC	07810 522515
Lyme Regis Power BC	01297 443788
Lyme Regis SC	01297 442373
Lymington Town SC	0159 674514
Lympstone SC Exeter	01395 278792
Madoc YC Porthmadog	01766 512976
Malahide YC	+353 1 845 3372
Maldon Little Ship Club	01621 854139
Manx Sailing & CC Ramsey	01624 813494
Marchwood YC	023 80666141
Margate YC	01843 292602
Marina BC Pwllheli	01758 612271
Maryport YC	01228 560865
Mayflower SC Plymouth	01752 662526
Mayo SC (Rosmoney) Rosmoney	+353 98 27772
Medway YC Rochester	01634 718399
Menai Bridge BC Beaumaris	01248 810583
Mengham Rythe SC Hayling Island	023 92463337
Merioneth YC Barmouth	01341 280000
Monkstone Cruising and SC Swansea	01792 812229
Montrose SC Montrose	01674 672554
Mumbles YC Swansea	01792 369321
Mylor YC Falmouth	01326 374391
Nairn SC	01667 453897
National YC Dun Laoghaire	+353 1 280 5725
Netley SC Netley	023 80454272
New Quay YC Aberdovey	01545 560516
Newhaven & Seaford SC Seaford	01323 890077
Newport and Uskmouth SC Cardiff	01633 271417
Newtownards SC	028 9181 3426
Neyland YC	01646 600267
North Devon YC Bideford	01271 861390
North Fambridge Yacht Centre	01621 740370
North Haven YC Poole	01202 708830
North of England Yachting Association Kirkwall	01856 872331
North Sunderland Marine Club Sunderland	01665 721231
North Wales CC Conwy	01492 593481
North West Venturers YC (Beaumaris) Beaumaris	0161 2921943
Oban SC Ledaig by Oban	01631 563999
Orford SC Woodbridge	01394 450997
Orkney SC Kirkwall	01856 872331
Orwell YC Ipswich	01473 602288
Oulton Broad Yacht Station	01502 574946
Ouse Amateur SC Kings Lynn	01553 772239
Paignton SC Paignton	01803 525817
Parkstone YC Poole	01202 743610
Peel Sailing and CC Peel	01624 842390
Pembroke Haven YC	01646 684403
Pembrokeshire YC Milford Haven	01646 692799
Penarth YC	029 20708196
Pentland Firth YC Thurso	01847 891803
Penzance YC	01736 364989
Peterhead SC Ellon	01779 75527
Pin Mill SC Woodbridge	01394 780271
Plym YC Plymouth	01752 404991
Poolbeg YC	+353 1 660 4681
Poole YC	01202 672687
Porlock Weir SC Watchet	01643 862702
Port Edgar YC Penicuik	0131 657 2854
Port Navas YC Falmouth	01326 340065
Port of Falmouth Sailing Association Falmouth	01326 372927
Portchester SC Portchester	023 9237 6375
Porthcawl Harbour BC Swansea	01656 655935
Porthmadog SC Porthmadog	01766 513546
Portrush YC Portrush	028 7082 3932
Portsmouth SC	02392 820596
Prestwick SC Prestwick	01292 671117
Pwllheli SC Pwllheli	01758 613343
Queenborough YC Queenborough	01795 663955
Quoile YC Downpatrick	028 44 612266
River Towy BC Tenby	01267 241755
RAFYC	023 80452208
Redclyffe YC Poole	01929 557227
Restronguet SC Falmouth	01326 374536
Ribble CC Lytham St Anne's	01253 739983
River Wyre YC	01253 811948
RNSA (Plymouth)	01752 55123/83
Rochester CC	01634 841350
Rock Sailing and Water Ski Club Wadebridge	01208 862431
Royal Dart YC Dartmouth	01803 752496
Royal Motor YC Poole	01202 707227
Royal Anglesey YC (Beaumaris) Anglesey	01248 810295

Royal Burnham YC Burnham-on-Crouch 01621 782044
Royal Channel Islands YC (Jersey) St Aubin 01534 745783
Royal Cinque Ports YC Dover 01304 206262
Royal Corinthian YC (Burnham-on-Crouch) Burnham-on-Crouch 01621 782105
Royal Corinthian YC (Cowes) Cowes 01983 293581
Royal Cork YC Crosshaven +353 214 831023
Royal Cornwall YC (RCYC) Falmouth 01326 312126
Royal Dorset YC Weymouth 01305 786258
Royal Forth YC Edinburgh 0131 552 3006
Royal Fowey YC Fowey 01726 833573
Royal Gourock YC 01475 632983
Royal Highland YC Nairn 01667 493855
Royal Irish YC Dun Laoghaire +353 1 280 9452
Royal London YC Isle of Wight 019 83299727
Royal Lymington YC 01590 672677
Royal Mersey YC Birkenhead 0151 645 3204
Royal Motor YC Poole 01202 707227
Royal Naval Club and Royal Albert YC Portsmouth 023 9282 5924
Royal Naval Sailing Association Gosport 023 9252 1100
Royal Norfolk & Suffolk YC Lowestoft 01502 566726
Royal North of Ireland YC 028 90 428041
Royal Northern and Clyde YC Rhu 01436 820322
Royal Northumberland YC Blyth 01670 353636
Royal Plymouth Corinthian YC Plymouth 01752 664327
Royal Scottish Motor YC 0141 881 1024
Royal Solent YC Yarmouth 01983 760256
Royal Southampton YC Southampton 023 8022 3352
Royal Southern YC Southampton 023 8045 0300
Royal St George YC Dun Laoghaire +353 1 280 1811
Royal Tay YC Dundee 01382 477133
Royal Temple YC Ramsgate 01843 591766
Royal Torbay YC Torquay 01803 292006
Royal Ulster YC Bangor 028 91 270568
Royal Victoria YC Fishbourne 01983 882325
Royal Welsh YC (Caernarfon) Caernarfon 01286 672599
Royal Welsh YC Aernarfon 01286 672599
Royal Western YC Plymouth 01752 226299
Royal Yacht Squadron Isle of Wight 01983 292191
Royal Yorkshire YC Bridlington 01262 672041
Rye Harbour SC 01797 223136
Salcombe YC 01548 842593
Saltash SC 01752 845988
Scalloway BC Lerwick 01595 880409
Scarborough YC 01723 373821
Schull SC +353 28 37352
Scillonian Sailing and BC St Mary's 01720 277229
Seasalter SC Whitstable 07773 189943
Seaview YC Isle of Wight 01983 613268
Shoreham SC Henfield 01273 453078
Skerries SC Carlingdford Lough +353 1 849 1233
Slaughden SC Duxford 01728 689036
Sligo YC Sligo +353 71 77168
Solva Boat Owners Association Fishguard 01437 721538
Solway YC Kirkdudbright 01556 620312
South Caernavonshire YC Abersoch 01758 712338
South Cork SC +353 28 36383
South Devon Sailing School Newton Abbot 01626 52352
South Gare Marine Club - Sail Section Middlesbrough 01642 505630
South Shields SC 0191 456 5821
South Woodham Ferrers YC Chelmsford 01245 325391
Southampton SC 023 8044 6575
Southwold SC 01986 784225
Sovereign Harbour YC Eastbourne 01323 470888
St Helier YC 01534 721307/32229
St Mawes SC 01326 270686
Starcross Fishing & CC (River Exe) Starcross 01626 891996
Starcross YC Exeter 01626 890470
Stoke SC Ipswich 01473 624989
Stornoway SC 01851 705412
Stour SC 01206 393924
Strangford Lough YC Newtownards 028 97 541202
Strangford SC Downpatrick 028 4488 1404
Strood YC Aylesford 01634 718261
Sunderland YC 0191 567 5133
Sunsail Portsmouth 023 92222224
Sussex YC Shoreham-by-Sea 01273 464868
Swanage SC 01929 422987
Swansea Yacht & Sub-Aqua Club Swansea 01792 469096
Tamar River SC Plymouth 01752 362741
Tarbert Lochfyne YC 01880 820376
Tay Corinthian BC Dundee 01382 553534
Tay YCs Association 01738 621860
Tees & Hartlepool YC 01429 233423
Tees SC Aycliffe Village 01429 265400
Teifi BC - Cardigan Bay Fishguard 01239 613846
Teign Corinthian YC Teignmouth 01626 777699
Tenby SC 01834 842762
Tenby YC 01834 842762
Thames Estuary YC 01702 345967
Thorney Island SC 01243 371731
Thorpe Bay YC 01702 587563
Thurrock YC Grays 01375 373720
Tollesbury CC 01621 869561
Topsham SC 01392 877524
Torpoint Mosquito SC - Plymouth 01752 812508
Tralee SC +353 66 7136119
Troon CC 01292 311190
Troon YC 01292 315315

Tudor SC Portsmouth 02392 662002
Tynemouth SC
Newcastle upon Tyne 0191 2572167
Up River YC
Hullbridge 01702 231654
Upnor SC 01634 718043
Vanguard SC
Workington 01228 674238
Wakering YC
Rochford 01702 530926
Waldringfield SC
Woodbridge 01394 283347
Walls Regatta Club
Lerwick 01595 809273
Walton & Frinton YC
Walton-on-the-Naze 01255 675526
Warrenpoint BC 028 4175 2137
Warsash SC Soton 01489 583575
Watchet Boat Owner Association
Watchet 01984 633736
Waterford Harbour SC
Dunmore East +353 51 383389
Watermouth YC
Watchet 01271 865048
Wear Boating Association
0191 567 5313
Wells SC
Wells-next-the-sea 01328 711190
West Kirby SC 0151 625 5579
West Mersea YC
Colchester 01206 382947
Western Isles YC 01688 302371
Western YC
Kilrush +353 87 2262885
Weston Bay YC
Portishead 07867 966429
Weston CC Soton 02380 466790
Weston SC Soton 02380 452527
Wexford HBC +353 53 22039
Weymouth SC 01305 785481
Whitby YC 07786 289393
Whitstable YC 01227 272942
Wicklow SC +353 404 67526
Witham SC Boston 01205 363598
Wivenhoe SC
Colchester 01206 822132
Woodbridge CC 01394 386737
Wormit BC 01382 553878
Yarmouth SC 01983 760270
Yealm YC
Newton Ferrers 01752 872291
Youghal Sailing Club +353 24 92447

YACHT DESIGNERS

Cannell & Associates, David M
Wivenhoe 01206 823337
Clarke Designs LLP, Owen
Dartmouth 01803 770495
Giles Naval Architects, Laurent
Lymington 01590 641777
Harvey Design, Ray
Barton on Sea 01425 613492
Jones Yacht Design, Stephen
Warsash 01489 576439
Wharram Designs, James
Truro 01872 864792
Wolstenholme Yacht Design
Coltishall 01603 737024

YACHT MANAGEMENT

Barbican Yacht Agency Ltd
Plymouth 01752 228855
Coastal Leisure Ltd
Southampton 023 8033 2222
O'Sullivan Boat Management
Dun Laoghaire +353 86 829 6625
Swanwick Yacht Surveyors
Swanwick 01489 564822
Amble Boat Company Ltd
Amble 01665 710267

YACHT VALETING

Autogleam
Lymington 0800 074 4672
Blackwell, Craig
Co Meath +353 87 677 9605
Bright 'N' Clean
South Coast 01273 604080
Clean It All
Nr Brixham 01803 844564
Kip Marina Inverkip 01475 521485
Mainstay Yacht Maintenance
Dartmouth 01803 839076
Mobile Yacht Maintenance
07900 148806
Shipshape Hayling Is 023 9232 4500